2D Materials

Edited by Chatchawal Wongchoosuk
and Yotsarayuth Seekaew

Published in London, United Kingdom

IntechOpen

Supporting open minds since 2005

2D Materials
http://dx.doi.org/10.5772/intechopen.78528
Edited by Chatchawal Wongchoosuk and Yotsarayuth Seekaew

Contributors
Bruno Chandrasekar L, S Nagarajan, Marimuthu Karunakaran, T. Daniel Thangadurai, Manuel Ramos, John Nogan, Claudia A. Rodriguez Gonzalez, José Luis Enríquez-Carrejo, José Mireles, Jr., Abel Hurtado-Macias, Carlos Ornelas, Roberto Carlos Ambrosio-Lázaro, Manuela Ortiz-Diaz, Torben Boll, Delphine Chassaing, Martin Heilmaier, Konstantin Zhuravlev, Yurij Galitsyn, Vladimir Mansurov, Timur Malin, Sergey Teys, Béla Pécz, Ildikó Cora, Godfrey Gumbs, Po-Hsin Shin, Thi Nga Do, Dipendra Dahal, Prabhakar Misra, Daniel Casimir, Iman Ahmed, Raul Garcia-Sanchez, Hawazin Alghamdi, Chatchawal Wongchoosuk

Notice
Statements and opinions expressed in the chapters are these of the individual contributors and not necessarily those of the editors or publisher. No responsibility is accepted for the accuracy of information contained in the published chapters. The publisher assumes no responsibility for any damage or injury to persons or property arising out of the use of any materials, instructions, methods or ideas contained in the book.

First published in London, United Kingdom, 2019 by IntechOpen
IntechOpen is the global imprint of INTECHOPEN LIMITED, registered in England and Wales, registration number: 11086078, The Shard, 25th floor, 32 London Bridge Street London, SE19SG – United Kingdom
Printed in Croatia

British Library Cataloguing-in-Publication Data
A catalogue record for this book is available from the British Library

Additional hard and PDF copies can be obtained from orders@intechopen.com

2D Materials
Edited by Chatchawal Wongchoosuk and Yotsarayuth Seekaew
p. cm.
Print ISBN 978-1-83962-262-5
Online ISBN 978-1-83962-263-2
eBook (PDF) ISBN 978-1-83962-264-9

We are IntechOpen,
the world's leading publisher of
Open Access books
Built by scientists, for scientists

4,300+
Open access books available

117,000+
International authors and editors

130M+
Downloads

Our authors are among the

151
Countries delivered to

Top 1%
most cited scientists

12.2%
Contributors from top 500 universities

CLARIVATE ANALYTICS
BOOK
CITATION
INDEX
INDEXED

WEB OF SCIENCE™

Selection of our books indexed in the Book Citation Index
in Web of Science™ Core Collection (BKCI)

Interested in publishing with us?
Contact book.department@intechopen.com

Numbers displayed above are based on latest data collected.
For more information visit www.intechopen.com

Meet the editors

Chatchawal Wongchoosuk obtained a BSc in Physics from Prince Songkla University, Thailand, in 2005, and PhD and MSc degrees from Mahidol University, Thailand, in 2007 and 2011, respectively. Currently, he is a faculty member of Kasetsart University, Bangkok. He is a specialist in development of smart sensors for food, agricultural and environmental applications. He has received more than 20 research awards. Dr. Wongchoosuk has published several dozens of articles in reputed journals and proceedings. He has served as a reviewer for more than 22 ISI journals. His research interests include modern nanoscience and nanotechnology ranging from theoretical modeling to fabrication of intelligent nanodevices and systems such as hybrid gas sensors, electronic noses, digital farms, smart sensors, printed electronics and flexible electronics.

Yotsarayuth Seekaew obtained a BSc, MSc and PhD in Physics from Kasetsart University, Thailand, in 2010, 2014 and 2019, respectively. Presently, he is a researcher at the Laboratory for Multiscale Innovative Technologies, Department of Physics, Faculty of Science, Kasetsart University. His research interests include graphene-based gas sensors, polymer sensors, electroluminescent sensing applications and printable sensors.

Contents

Preface

Two-dimensional (2D) materials have attracted a great deal of attention in recent years due to their potential applications in gas/chemical sensors, healthcare monitoring, biomedicine, electronic skin, wearable sensing technology, flat panel displays, optoelectronics, photodetectors, catalysis, electrochemical sensing, bio sensing, water/air purification, batteries, fuel cells and advanced electronic devices. One of the most popular 2D nanomaterials in this era is graphene, which has unique properties such as large specific surface area, high electrical conductivity, excellent electron transfer rate and high mechanical strength. However, it is not only 2D graphene that has been widely applied in a large variety of potential applications but also other 2D materials such as boron nitrides, molybdenum disulfide, black phosphorus and metal oxide nanosheets, all of which open up new opportunities for future devices.

This book focuses on models and theoretical backgrounds, important properties, characterizations and applications of current, popular 2D materials such as graphene, silicon nitride, aluminum nitride, ZnO thin films, phosphorene and molybdenum disulfide. Chapter 1 presents an overview, synthesis methods and applications of popular 2D materials. Chapter 2 focuses on properties and characterizations of graphene, graphite and graphene nanoplatelets via Raman spectroscopy. Chapter 3 provides an insight into the properties of undoped and doped ZnO thin films. Chapter 4 describes structures, kinetics and thermodynamics of 2D silicon nitride and aluminum nitride. Chapter 5 focuses on structure, electronic properties, polarizability and dielectric function of 2D phosphorene based on theoretical approaches via the tight-binding model. Chapter 6 demonstrates synthesis, characterizations, and mechanical and electrical properties of molybdenum disulfide as well as its photovoltaic applications.

We would like to express our deep appreciation to all contributors who are experts in their respective research fields. It should be emphasized that all chapters have been submitted to re-review and revision in order to improve their presentation with several interactions between editors, authors and publisher. We hope this book will be a useful tool and provide inspiration and motivation to interested readers for further developments in the field.

Asst. Prof. Dr. Chatchawal Wongchoosuk and Dr. Yotsarayuth Seekaew
Department of Physics, Faculty of Science,
Kasetsart University,
Thailand

Chapter 1

Introductory Chapter: 2D Materials

Yotsarayuth Seekaew and Chatchawal Wongchoosuk

1. Overview

Two-dimensional (2D) materials are a class of nanomaterials that have two dimensions (XY plane) outside of the nanometric size range and atomic-scale thicknesses (Z dimension). The first well-known 2D material is graphene consisting of a single layer of carbon atoms arranged in a hexagonal lattice. To compare with 0D material (fullerene) and 1D material (carbon nanotube), the researches related to 2D material (graphene) have grown up quickly over other carbon allotropes as shown in **Figure 1**. Based on Scopus database (search by keyword "graphene" on March 18, 2019), publications on graphene increased from 3772 papers in 2010 to 21,439 papers in 2018. The total number of graphene-related publications is 132,628 documents. However, it is not only 2D graphene that has been widely applied in a large variety of potential applications but also other 2D materials such as tungsten disulfide, molybdenum disulfide, and silicon nitride open up new opportunities for the future devices. In this chapter, synthesis and applications of these 2D materials have been introduced and presented in brief.

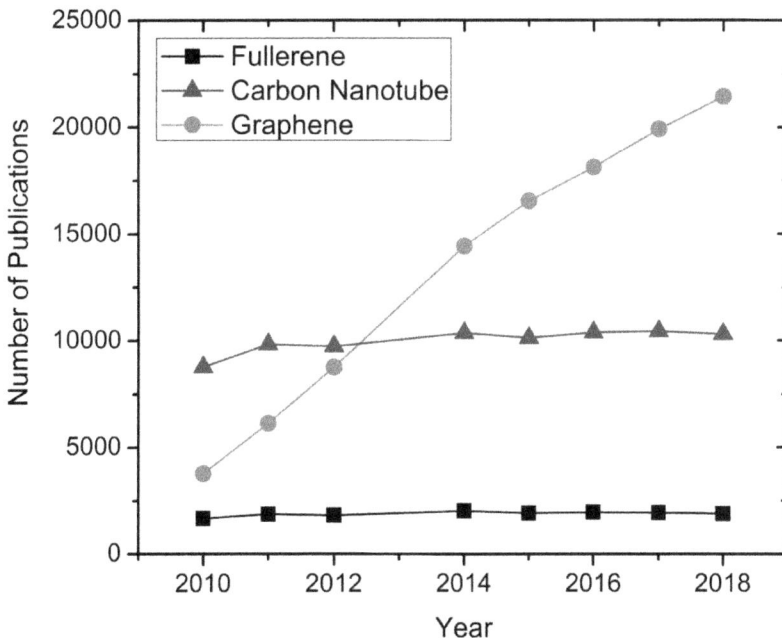

Figure 1.
Number of publications versus publication years based on Scopus database (search by keyword "fullerene," "carbon nanotube," and "graphene" on March 18, 2019).

2. Synthesis methods of 2D materials

2.1 Graphene

Graphene can be synthesized by several methods depending on the required quality and quantity. (I) Chemical exfoliation method by modified Hummers method [1] is one of the popular methods for graphene oxide growth based on suitable oxidizing agents from graphite oxide. This method offers a large amount of graphene products and is of low cost. (II) Electrochemical exfoliation method is based on formation of graphene product from graphite rod or highly orientated pyrolytic graphite (HOPG) by using electricity for exfoliation of the graphite rod or HOPG immersed into electrolyte solutions [2]. (III) Chemical vapor deposition (CVD) method provides high-quality graphene products with controllable graphene layers over a large-scale area [3, 4]. Usually, methane (CH_4) and acetylene (C_2H_2) were used as carbon source for graphene growths on copper (Cu) or nickel (Ni) foam under high temperature around 1000°C.

2.2 Tungsten disulfide (WS_2)

The synthesis of tungsten disulfide (WS_2) can be done by three main methods, namely hydrothermal method, atomic layer deposition (ALD), and CVD. A simple hydrothermal method was used to form WS_2/C composite using $Na_2WO_4 \cdot 2H_2O$ and CH_3CSNH_2 as raw materials, polyethylene glycol as dispersant, and glucose as the carbon source under annealing at a low temperature in argon atmosphere [5]. ALD was employed to form mono-, bi-, and multilayer WS_2 nanosheets by controlling the number of cycles of ALD WO_3 with plasma enhancement using WH_2 (iPrCp)$_2$ and oxygen [6]. The synthesis process of large-area WS_2 films based on CVD can be described as follows [7]: (I) the Na_2WO_4 precursor coated on SiO_2/Si substrate was loaded into quartz tube of CVD process. (II) Argon was flowed into the quartz tube until temperature reached 850°C. (III) A liquid phase of dimethyl disulfide ((CH_3)$_2S_2$, DMDS) was introduced with a bubbling system for 30 min to form the WS_2 film.

2.3 Molybdenum disulfide (MoS_2)

MoS_2 can be synthesized by using mechanical and chemical methods. For example, single-layer and multilayer MoS_2 nanosheets were formed by using adhesive Scotch tape from transition metal dichalcogenide (TMD) materials [8]. MoS_2 nanosheets were synthesized from $NaBH_4$ as a reductant by chemical exfoliation [9] and liquid-phase exfoliation method with N-methyl-2-pyrrolidone (NMP) solvents [10]. Moreover, MoS_2 can be prepared via hydrothermal method, ALD, and CVD. For example, MoS_2 nanospheres were formed with $Na_2MoO_4 \cdot 2H_2O$ dissolved in DDW by hydrothermal method [11]. MoS_2 atomic layers were synthesized from MoO_3 and pure sulfur in a vapor-phase-deposition process with a reaction temperature of 850°C [12]. Based on CVD, the synthesis of MoS_2 was prepared from high purity MoO_3 powder and S powder in two separate Al_2O_3 crucibles and placed into quartz tube of CVD process. The SiO_2/Si substrates were faced down and placed on the crucible of MoO_3 powder together with annealing at 650°C for 15 min and N_2 flow (1 sccm) at ambient to obtain 2D-MoS_2 on Si substrates [13].

2.4 Silicon nitride (Si_3N_4)

Si_3N_4 has been widely synthesized by using carbothermal and nitriding reactions. For example, SiO_2/C mixture on alumina boat was placed in a high temperature tubular furnace with a flow rate of nitrogen and hydrogen under optimal condition to promote the formation of Si_3N_4 [14]. Fe-Si_3N_4 composite was also prepared by $FeSi_{75}$ powder as a precursor under reaction of high purity nitrogen flow via flash combustion at a high temperature of 1450°C [15].

3. Applications of 2D materials

3.1 Graphene

Graphene has been widely used for various applications including energy storage, solar cells, and gas sensor. Abdelkader et al. [16] reported the fabrication of flexible printed graphene supercapacitor device for wearable electronics by using graphene oxide ink and a screen-printing technique. The supercapacitor device can give a capacitance as high as 2.5 mF cm^{-2} and maintain 95.6% in cyclic stability over 10,000 cycles. Shin et al. [17] reported the fabrication of graphene/porous silicon Schottky-type solar cells by doping with silver nanowires (AgNWs) into graphene/porous silicon nanocomposite. Moreover, graphene has been widely applied in sensing application. For example, graphene was combined with carbon nanotubes to form as the 3D carbon nanostructures or the pillared graphene structures for toluene-sensing applications at room temperature [18]. We reported fabrication of various layer graphene gas sensors for NO_2 detection and investigated the layer effect of graphene to NO_2 detection. We found that bilayer graphene gas sensor exhibited the highest response and highest sensitivity to NO_2 at room temperature due to accessible active surface area and unique band structure of bilayer graphene [3]. Very recently, we demonstrated a new type of graphene gas sensor based on AC electroluminescent (EL) principle [4]. This device can monitor carbon dioxide (CO_2) at room temperature via changing El emission upon CO_2 gas concentration. Advantage of our graphene-based electroluminescent gas sensor over typical current gas sensor is to directly integrate with a smart phone via light sensor without any modification of smart phone hardware.

3.2 Tungsten disulfide (WS_2)

WS_2 nanoflakes were used for lithium ion battery applications. They showed reversible capacity of 680 mA h/g and 86.2% of the initial capacity after 20 cycles [19]. Pawbake et al. reported that WS_2 nanoparticle was used for photodetector and humidity sensing applications [20]. It was found that the WS_2 nanoparticle-based humidity sensor exhibited sensitivity of 469%, response time of ~12 s, and recovery time of ~13 s. In case of based photodetection application, WS_2 showed a sensitivity of ~137% under white light illumination. The response and recovery times were ~51 and ~88 s, respectively [20].

3.3 Molybdenum disulfide (MoS_2)

MoS_2 have been extensively applied in sensor, optical, energy device, and electronics. For example, tactile sensor was fabricated from MoS_2 for electronic skin applications. MoS_2 owns its outstanding properties such as good optical transparency, mechanical flexibility, and high gauge factor compared with conventional strain gauges [21]. Wang et al. studied the conductivity and thermal stability of the MoS_2/polyaniline (PANI) nanocomposites with increasing the amount of MoS_2 for supercapacitor application. The results showed that the MoS_2/PANI of 38 wt% exhibited specific capacitance up to 390 F/g and retained capacitance of 86% over

1000 cycles [22]. MoS$_2$ was also synthesized to form hydrangea-like flowers or clusters comprising MoS$_2$ nanosheet for high-dielectric and electrical energy storage applications [23]. Moreover, Yin et al. synthesized the biocompatible nanoflowers between MoS$_2$ with polyethylene glycol (PEG) for antibacterial applications [24].

3.4 Silicon nitride (Si$_3$N$_4$)

Most applications of Si$_3$N$_4$ have been used in terms of the improvement of properties such as surface modulation for orthopedic applications [25] and biomedical applications [26]. Also, Si$_3$N$_4$ owns good optical properties. The Si$_3$N$_4$ was fabricated as photonic circuits to spectroscopic sensing [27]. The Si$_3$N$_4$ was used for nonlinear signal processing applications [28]. Furthermore, Si$_3$N$_4$ was microfabricated as the waveguides and grating couplers for new nanophotonic approach of light delivery for optogenetic applications [29].

4. Conclusion

In summary, the emerging 2D materials provide high impacts for science and advanced technologies. They own unique physical, optical, mechanical, and electrical properties. Therefore, 2D materials have become one of the hottest topics in this era due to their potential various applications such as gas/chemical sensors, healthcare monitoring, biomedicine, electronic skin, wearable sensing technology, flat panel displays, optoelectronics, photodetector, catalysis, electrochemical sensing, bio sensing, water/air purification, supercapacitor, batteries, fuel cells, and advanced electronics devices.

Acknowledgements

This work was supported by the Kasetsart University Research and Development Institute (KURDI). Y.S. acknowledges the Ph.D. Graduate Program Scholarship from the Graduate School, Kasetsart University and the National Research Council of Thailand (NRCT) as of fiscal year 2018.

Author details

Yotsarayuth Seekaew and Chatchawal Wongchoosuk*
Department of Physics, Faculty of Science, Kasetsart University, Bangkok, Thailand

*Address all correspondence to: chatchawal.w@ku.ac.th

IntechOpen

References

[1] Hummers WS Jr, Offeman RE. Preparation of graphite oxide. Journal of the American Chemical Society. 1958;**80**:1339-1339

[2] Rao KS, Senthilnathan J, Liu Y-F, Yoshimura M. Role of peroxide ions in formation of graphene nanosheets by electrochemical exfoliation of graphite. Scientific Reports. 2014;**4**:4237

[3] Seekaew Y, Phokharatkul D, Wisitsoraat A, Wongchoosuka C. Highly sensitive and selective room-temperature NO2 gas sensor based on bilayer transferred chemical vapor deposited graphene. Applied Surface Science. 2017;**404**:357-363

[4] Seekaew Y, Wongchoosuka C. A novel graphene-based electroluminescent gas sensor for carbon dioxide detection. Applied Surface Science. 2019;**479**:525-531

[5] Yuan Z, Jiang Q, Feng C, Chen X, Guo Z. Synthesis and performance of tungsten disulfide/carbon (WS_2/C) composite as anode material. Journal of Electronic Materials. 2018;**47**:251-260

[6] Song J-G, Park J, Lee W, Choi T, Jung H, Lee CW, et al. Layer-controlled, wafer-scale, and conformal synthesis of tungsten disulfide nanosheets using atomic layer deposition. ACS Nano. 2013;**12**:11333-11340

[7] Choi SH, Boandoh S, Lee YH, Lee JS, Park J-H, Kim SM, et al. Synthesis of large-area tungsten disulfide films on pre-reduced tungsten suboxide substrates. ACS Applied Materials & Interfaces. 2017;**9**:43021-43029

[8] Li H, Wu J, Yin Z, Zhang H. Preparation and applications of mechanically exfoliated single-layer and multilayer MoS_2 and WSe_2 nanosheets. Accounts of Chemical Research. 2014;**47**:1067-1075

[9] Guardia L, Paredes JI, Munuera JM, Villar-Rodil S, Ayan-Varela M, Martinez-Alonso A, et al. Chemically exfoliated MoS_2 nanosheets as an efficient catalyst for reduction reactions in the aqueous phase. ACS Applied Materials & Interfaces. 2014;**6**:21702-21710

[10] Gupta A, Arunachalam V, Vasudevan S. Liquid-phase exfoliation of MoS_2 nanosheets: The critical role of trace water. Journal of Physical Chemistry Letters. 2016;**7**:4884-4890

[11] Chung DY, Park SK, Chung YH, Yu SH, Lim DH, Jung N, et al. Edge-exposed MoS_2 nano-assembled structures as efficient electrocatalysts for hydrogen evolution reaction. Nanoscale. 2014;**6**:2131-2136

[12] Najmaei S, Liu Z, Zhou W, Zou X, Shi G, Lei S, et al. Vapour phase growth and grain boundary structure of molybdenum disulphide atomic layers. Nature Materials. 2013;**12**:754-759

[13] Lee YH, Zhang XQ, Zhang W, Chang MT, Lin CT, Chang KD, et al. Synthesis of large-area MoS_2 atomic layers with chemical vapor deposition. Advanced Materials. 2012;**24**:2320-2325

[14] Ortega A, Alcalá MD, Real C. Carbothermal synthesis of silicon nitride (Si_3N_4): Kinetics and diffusion mechanism. Journal of Materials Processing Technology. 2008;**195**:224-231

[15] Li B, Li G, Chen H, Chen J, Hou X, Li Y. Reaction and formation mechanism of $Fe-Si_3N_4$ composite prepared by flash combustion synthesis. Ceramics International. 2018;**44**:22777-22783

[16] Abdelkader AM, Karim N, Vallés C, Afroj S, Novoselov KS, Yeates SG. Ultraflexible and robust graphene supercapacitors printed on textiles for

wearable electronics applications. 2D Materials. 2017;**4**:035016

[17] Shin DH, Kim JH, Kim JH, Jang CW, Seo SW, Lee HS, et al. Graphene/porous silicon Schottky-junction solar cells. Journal of Alloys and Compounds. 2017;**715**:291-296

[18] Seekaew Y, Wisitsoraat A, Phokharatkul D, Wongchoosuk C. Room temperature toluene gas sensor based on TiO_2 nanoparticles decorated 3D graphene-carbon nanotube nanostructures. Sensors and Actuators B: Chemical. 2019;**279**:69-78

[19] Feng C, Huang L, Guo Z, Liu H. Synthesis of tungsten disulfide (WS_2) nanoflakes for lithium ion battery application. Electrochemistry Communications. 2007;**9**:119-122

[20] Pawbake AS, Waykar RG, Late DJ, Jadkar SR. Highly transparent wafer-scale synthesis of crystalline WS_2 nanoparticle thin film forphotodetector and humidity-sensing applications. ACS Applied Materials & Interfaces. 2016;**8**:3359-3365

[21] Park M, Park YJ, Chen X, Park Y-K, Kim M-S, Ahn J-H. MoS_2-based tactile sensor for electronic skin applications. Advanced Materials. 2016;**28**:2556-2562

[22] Wang J, Wu Z, Hu K, Chen X, Yin H. High conductivity graphene-like MoS_2/polyaniline nanocomposites and its application in supercapacitor. Journal of Alloys and Compounds. 2015;**619**:38-43

[23] Jia Q, Huang X, Wang G, Diao J, Jiang P. MoS_2 nanosheet superstructures based polymer composites for high-dielectric and electrical energy storage applications. Journal of Physical Chemistry C. 2016;**120**:10206-10214

[24] Yin W, Yu J, Lv F, Yan L, Zheng LR, Gu Z, et al. Functionalized nano-MoS_2 with peroxidase catalytic and near-infrared photothermal activities for safe and synergetic wound antibacterial applications. ACS Nano. 2016;**10**:11000-11011

[25] Bock RM, McEntire BJ, Bal BS, Rahaman MN, Boffelli M, Pezzotti G. Surface modulation of silicon nitride ceramics for orthopaedic applications. Acta Biomaterialia. 2015;**26**:318-330

[26] Zhao S, Xiao W, Rahaman MN, O'Brien D, Sampson JWS, Bal BS. Robocasting of silicon nitride with controllable shape and architecture for biomedical applications. International Journal of Applied Ceramic Technology. 2017;**14**:117-127

[27] Ananth Z et al. Silicon and silicon nitride photonic circuits for spectroscopic sensing on-a-chip. Photonics Research. 2015;**3**:47-59

[28] Lacava C, Stankovic S, Khokhar AZ, Bucio TD, Gardes FY, Reed GT, et al. Si-rich silicon nitride for nonlinear signal processing applications. Scientific Reports. 2017;**7**:22

[29] Shim E, Chen Y, Masmanidis S, Li M. Multisite silicon neural probes with integrated silicon nitride waveguides and gratings for optogenetic applications. Scientific Reports. 2016;**6**:22693

Chapter 2

Raman Spectroscopy of Graphene, Graphite and Graphene Nanoplatelets

Daniel Casimir, Hawazin Alghamdi, Iman Y. Ahmed,
Raul Garcia-Sanchez and Prabhakar Misra

Abstract

The theoretical simplicity of sp^2 carbons, owing to their having a single atomic type per unit cell, makes these materials excellent candidates in quantum chemical descriptions of vibrational and electronic energy levels. Theoretical discoveries, associated with sp^2 carbons, such as the Kohn anomaly, electron-phonon interactions, and other exciton-related effects, may be transferred to other potential 2D materials. The information derived from the unique Raman bands from a single layer of carbon atoms also helps in understanding the new physics associated with this material, as well as other two-dimensional materials. The following chapter describes our studies of the G, D, and G′ bands of graphene and graphite, and the characteristic information provided by each material. The G-band peak located at ~1586 cm^{-1}, common to all sp^2 carbons, has been used extensively by us in the estimation of thermal conductivity and thermal expansion characteristics of the sp^2 nanocarbon associated with single walled carbon nanotubes (SWCNT). Scanning electron microscope (SEM) images of functionalized graphene nanoplatelet aggregates doped with argon (A), carboxyl (B), oxygen (C), ammonia (D), fluorocarbon (E), and nitrogen (F), have also been recorded and analyzed using the Gwyddion software.

Keywords: Raman spectroscopy, 2-D materials, graphene, graphite, functionalized graphene nanoplatelets

1. Introduction

The elucidation of novel physics related to 2D electronic systems (2DES) has received wide recognition in the form of three Nobel Prizes in Physics in 1985 [1] (Klaus von Klitzing, Max Planck Institute, for the discovery of the integer Quantum Hall Effect), in 1998 [2] (Robert Laughlin, Stanford University, Horst Stormer, Columbia University, and Daniel Tsui, Princeton University, for the discovery of the fractional Quantum Hall Effect), and in 2010 [3] (Andre Geim and Konstantin Novoselov, University of Manchester, for ground-breaking experiments relating to the 2D material graphene).

Since graphene can be considered as the conceptual parent material for all other sp^2 nanocarbons, it is the first in our discussion of the two-dimensional characteristics obtainable via Raman spectroscopy. Graphene is a two-dimensional carbon nanomaterial with a single layer of sp^2-hybridized carbon atoms arranged in a crystalline structure of six-membered rings [4, 5]. **Figure 1** illustrates the hexagonal lattice of a perfectly flat graphene sheet and the resulting nanotube after it is rolled along the vector labeled C_h. The shaded portion of the nanotube in **Figure 1(b)** represents one unit cell of the resulting armchair nanotube in this case, and it results from rolling the initial planar sheet in **Figure 1(a)**, so that points **A** and **C** coincide with points **B** and **D**, respectively. C_h is known as the chiral vector and is constructed from the vector addition of the graphene basis vectors a_1 and a_2. The integer number of each of the basic lattice vectors used in the construction, n and m, designated for a_1 and a_2 respectively, is arbitrary with the only provision that $(0 \leq |m| \leq n)$. The Cartesian components of the lattice vectors a_1 and a_2 are $(a\sqrt{3}/2, a/2)$ and $(a\sqrt{3}/2, -a/2)$, respectively, where the quantity $a = a_{C-C}\sqrt{3} = 2.46$ Å. The quantity a_{C-C} is the bond length between two neighboring carbon atoms in the hexagonal lattice equal to 1.42 Å. The chiral vector C_h is usually written in terms of the two integers n and m as

$$C_h = n\,a_1 + m\,a_2 \tag{1}$$

and has a magnitude of

$$|C_h| = a\,\sqrt{n^2 + m^2 + nm} \tag{2}$$

which equals the carbon nanotube's circumference. In a fashion similar to applying the above rolling operation on the graphene unit cell in **Figure 1** for the construction of single walled carbon nanotubes, graphite can be described in terms of stacking multiple graphene layers one atop the other.

We have also investigated functionalized graphene nanoplatelets, which are comprised of platelet-shaped graphene sheets, identical to those found in SWCNT, but in planar form. Among the samples we used (functionalized oxygen, nitrogen, argon, ammonia, carboxyl and fluorocarbon) all have similar shape. Graphene nanoplatelet aggregates (aggregates of sub-micron platelets with a diameter of <2 μm and a thickness of a few nanometers) were identified, rather than individual nanoplatelets (STREM Data Sheets [6]). According to the manufacturer's (Graphene Supermarket™), structural analysis for fluorinated graphene nanoplatelets (GNP), the lateral dimensions of the wrinkled sheet-like outer surface of the GNP's is ~1–5 μm [6]. The quoted number of graphene layers was also around 37

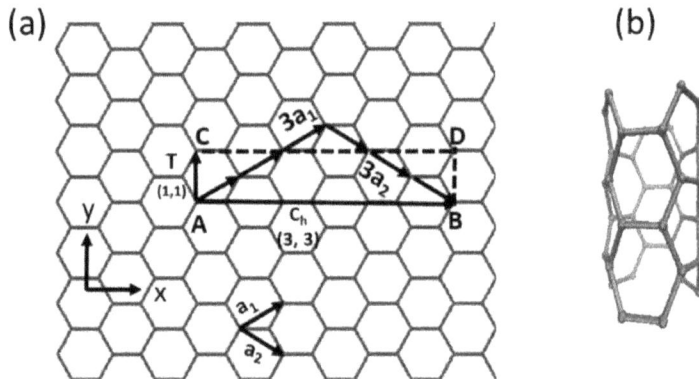

Figure 1.
(a) *Construction of a carbon nanotube based on the lattice vectors of a planar hexagonal graphene sheet.*
(b) *The resulting armchair (3, 3) carbon nanotube.*

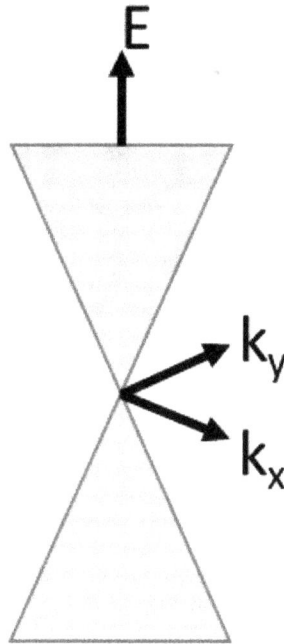

Figure 2.
The linear E vs. k dispersion of graphene near the Brillouin zone K-point (Dirac cone).

layers, based on SEM and TEM analysis [7]. This average number of layers is more than sufficient to consider these sub-micron sized stacked graphene nanoparticles as being in the multi-layer graphene category, and not in the few layer category. This latter property of GNP's is also mentioned in the concluding remarks, in connection with their two-dimensional classification.

Arguably one of the most striking displays of the two-dimensional nature of graphene is related to its electronic structure; specifically, the behavior of its electron/hole carriers [4]. The resulting dispersion (E vs. **k**) relation of the graphene band structure forces us to rely on Dirac's relativistic wave equation, instead of Schrodinger's equation, to describe the particle dynamics [4]. Therefore, the charge carriers are treated as relativistic massless quantities moving essentially at the speed of light [4].

Figure 2 shows one of the six "Dirac cones," which are one of the highly symmetrized K point locations in graphene's Brillouin zone, as shown in **Figure 3**, where the valence and conduction bands touch one another [4]. The experimental verification of this linear dispersion for energies centered at and near the Dirac cone as expressed in Eq. (3) has also been accomplished by various spectroscopic methods [8].

$$E\left(\vec{k}\right)^{\pm} = \pm \hbar v_F |\vec{k}| \tag{3}$$

Another significant electronic structural quantity that expresses the two-dimensional nature of graphene is the density of states (DOS), $g(\mathbf{E})$, which as its name suggests, gives the density of mobile charge carriers that are available at some temperature T [4]. Unsurprisingly, this quantity also varies linearly, only this time with the energy E, as expressed in Eq. (4) [4].

$$g(E) = \frac{2}{\pi (\hbar v_F)^2} |E| \tag{4}$$

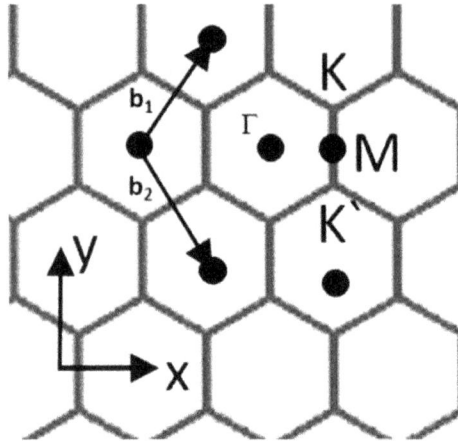

Figure 3.
Graphene's reciprocal space lattice shown with reciprocal lattice vectors b1, and b2. The first Brillouin zone is the region labeled by Γ. Also shown are the six high symmetry regions, Γ, K, M, and K'.

The 2D nature of graphene plays a direct role in this result, sine g(E), which gives the number of available states within the energy interval E and E + dE, is defined in two-dimensions in terms of the ratio of an element of area dA in k-space per unit wave-vector k.

2. Experimental

Figure 4 shows the primary instrument used to record most of the Raman spectra; it was a DXR SmartRaman spectrometer from Thermoelectron (that uses 780, 532, and 455 nm laser sources). The first wavelength (780 nm) was used for the bulk of the recorded spectra and utilized a high brightness laser of the single mode diode (as does the 532 nm light source), while the 455 nm source is a diode-pumped solid-state laser. Using the 180° geometry, after focusing the laser beam on the sample, the backscattered radiation from the sample enters the spectrometer via a collection lens, and the Stokes-shifted Raman spectrum was recorded as read by the CCD detector using the correct Rayleigh filter and automated entrance slit selections. A full range grating was used with the triplet spectrograph.

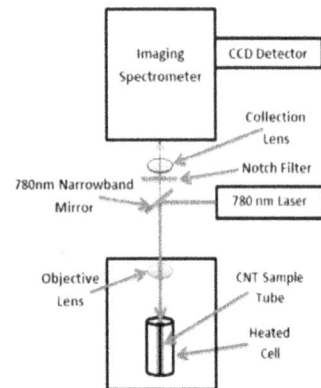

Figure 4.
DXR SmartRaman spectrometer (left) and DXR schematic (right).

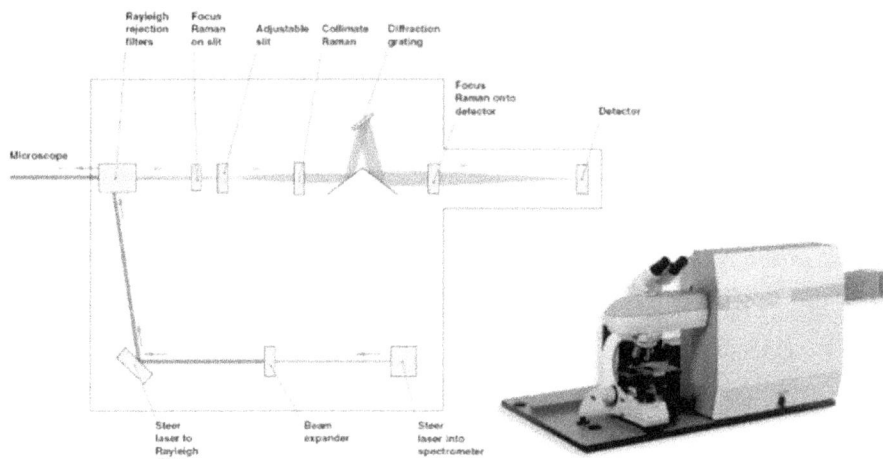

Figure 5.
InVia Raman spectrometer schematic and instrument image from the Renishaw manual.

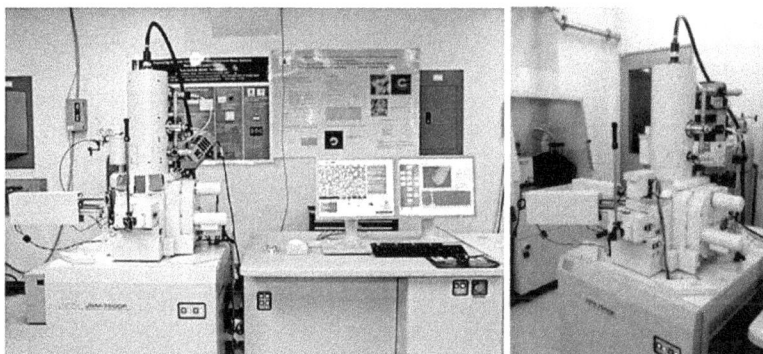

Figure 6.
Two views of the JEOL JSM-7600F scanning electron microscope (SEM) setup.

The Renishaw inVia Raman spectrometer (**Figure 5**) uses a 532-nm laser source and was employed to record the Raman spectra of CVD graphene and the functionalized graphene nanoplatelets. It consists of a microscope to shine light on the sample and collecting the scattered light, after filtering all the light except for the tiny fraction that has been Raman scattered, together with a diffraction grating for splitting the Raman scattered light into its component wavelengths, and a CCD camera for final detection of the Raman spectrum.

Graphene nanoplatelets (GNP) are usually produced by the intercalation of graphite through various means, followed by an acid purification process, and further exfoliation of the initial GNP flakes [9]. Besides intercalation, irradiation with microwaves, or extreme heating is also sometimes used to produce GNP from the host graphite source [10].

Images of functionalized graphene nanoplatelets were taken with the JEOL JSM-7600F scanning electron microscope (SEM) shown in **Figure 6**. The secondary electron detector on the SEM uses an EMI current of 138.20 nA. Beam current employed had a range of 1 pA to 200 nA. The JEOL JSM-7600F SEM contains a large variety of detectors that can be used on specimen samples up to 200 mm in diameter. Various magnifications were selected when appropriate to accurately display sample structure; SEM magnification ranges between 25 and 1,000,000×. The modular software program Gwyddion was used to generate three-dimensional visualization of the nanoplatelet aggregate structures.

3. Results

The structural simplicity of graphene is also exhibited in its Raman spectrum in contrast to its other fullerene relatives [11]. The two prominent bands located at 1580 and 2700 cm^{-1} are customarily called the G and G′-bands, respectively [11]. The high energy first order G-band has been identified with the intra-planar stretching modes of the strongly connected σ-bonded carbons [5]. The G′-band at 2700 cm^{-1} is attributed to a second order Raman scattering event with the phonon wave vector $q \neq 0$ [5]. **Figure 7** shows both bands obtained from a graphene sample on a nickel substrate.

Discerning the two-dimensional nature of graphene can be accomplished by contrasting the G′-band features of graphite and the former material [11]. First, the relative intensities between the G and G′ bands are different for graphene and its macroscopic relative graphite. In the case of graphene, the G′-band has a greater intensity than the G-band, which is the case for G-bands illustrated in **Figures 7** and 8. The G′-band of graphite is also shifted to a higher frequency compared to that of graphene [11]. Thirdly, the overall shape of the G′-band is usually more uniform compared to that of graphite, usually requiring a single Lorentzian to be fitted [11]. This last effect especially arises due to interactions among the multiple layers of graphite [11]. The Raman spectrum of the graphene sample was recently collected on an aged sample, and the degradation and contamination of this extremely thin material over time may be responsible for our Raman spectra of graphene and graphite only satisfying the first of these three criteria convincingly.

Not only the dimensionality, or number of layers present can be obtained via the Raman bands of graphene or graphite, but the average lateral characteristic size can of the graphene layers in the beam spot can also be determined. This was initially

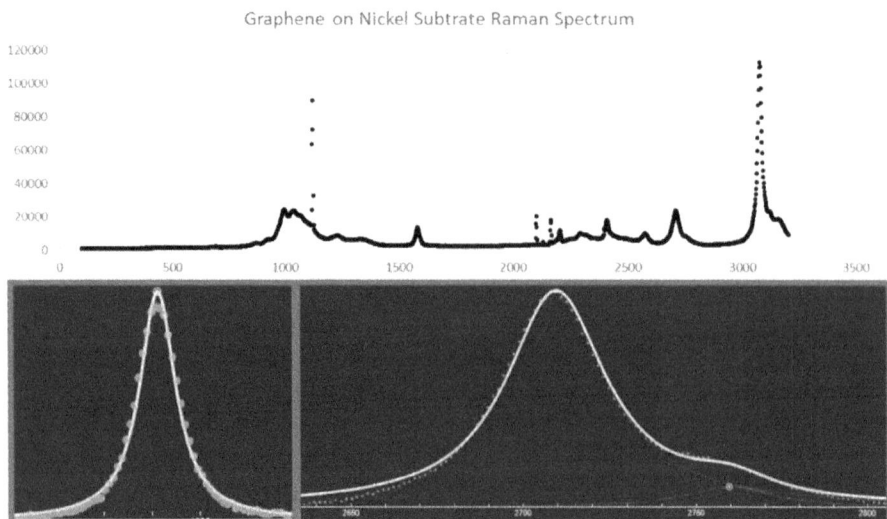

Figure 7.
(Top) Raman spectrum of CVD graphene on nickel substrate collected using 514 nm laser excitation. (Bottom left) G-band and Lorentzian (1582.4 cm^{-1}, height: 11,655.9). (Bottom right) G′-band and Lorentzians (2709.6 cm^{-1}, 2759.6 cm^{-1}, heights: 19,669.9, 1856.6).

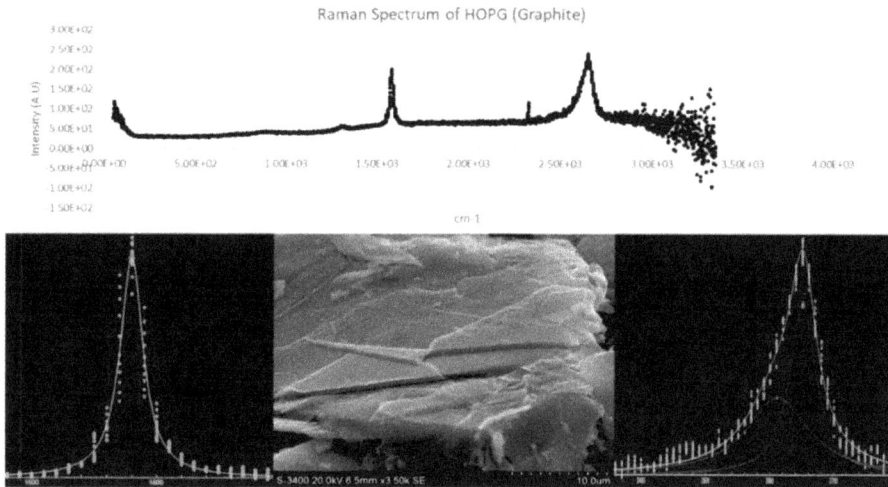

Figure 8.
(Top) Raman spectrum of HOPG graphite at excitation of 780 nm. (Bottom left) G-band and Lorentzian (1579.7 cm^{-1}, height: 128.9); (Bottom middle) SEM image of HOPG sample; (Bottom right) G'-band and Lorentzians (2611.8 cm^{-1}, 2651 cm^{-1}, heights: 56.1, 123.1).

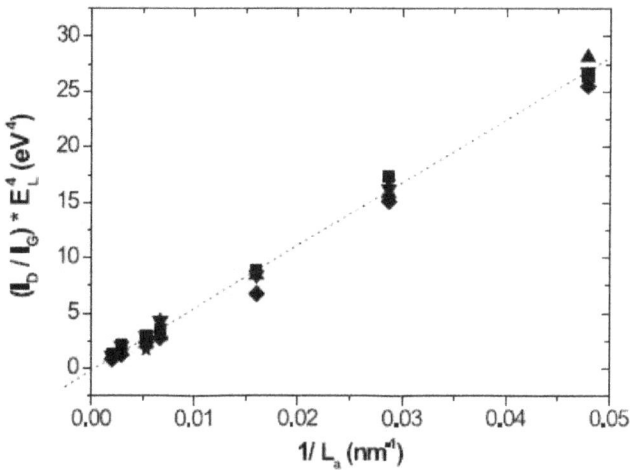

Figure 9.
Plot of $(I_D/I_G)*(E_L)^4$ vs. $1/L_a$. E_L is the laser excitation energy in eV, I_D and I_G are the D and G band intensities, respectively, and L_a is the characteristic lateral size of the graphene layer. Adapted from [5].

discovered by Tuinstra and Koenig, who correctly deduced that the intensity ratio of the D and G-bands varies directly with the characteristic size La of the planar graphite crystallites [12]. Further work done by Cancado et al. [13], expanded on Tuinstra and Koenig's work, by demonstrating the excitation energy dependence of the proportionality factor in the original relation as shown in **Figure 9** and as expressed in Eq. (5).

$$L_a = \left(2.4 \times 10^{-10}\right) \lambda^4 \left(\frac{I_D}{I_G}\right)^{-1} \tag{5}$$

For the graphene sample in **Figure 7** with a D-band intensity of 2719.7 and the graphite sample in **Figure 8** with a D-band intensity of 8.6, the respective L_a values are 71.8 nm and 1.3 μm according to Eq. (5) (**Table 1**).

Element	x average (µm)	y average (µm)	z average (µm)
Argon	4.8	3.9	0.50
Carboxyl	4.3	4.5	0.57
Oxygen	4.7	4.3	0.90
Ammonia	4.4	3.7	0.64
Fluorocarbon	5.0	3.6	0.55
Nitrogen	6.7	6.5	0.91

Table 1.
Average x, y, z axis measurements of functionalized graphene nanoplatelet aggregates.

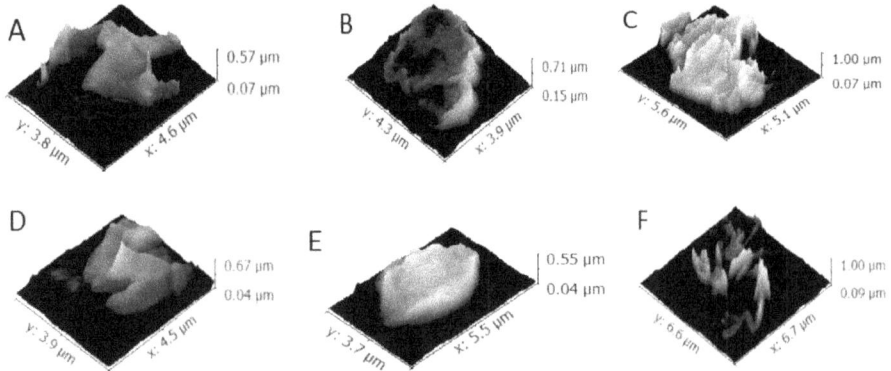

Figure 10.
3D view of SEM data of functionalized graphene nanoplatelet aggregates doped with argon (A), carboxyl (B), oxygen (C), ammonia (D), fluorocarbon (E), and nitrogen (F), respectively, using Gwyddion software.

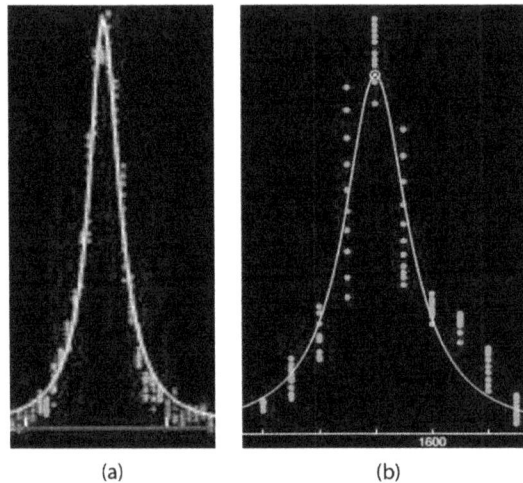

(a)   (b)

Figure 11.
(a) D-band on left (Intensity: 33.5, Center: 1312.4 cm^{-1}), (b) G-band on right (Intensity: 62.5, Center: 1580.2 cm^{-1}) for graphene nanoplatelets (ammonia) with 780 nm excitation and using fityk peak fitting software [14].

The observed sub-micron size of the platelets obtained mentioned above, and imaged in **Figure 10**, was also verified via Raman spectroscopy, based on the use of Eq. (3) for the graphene nanoplatelets functionalized with Ammonia, whose D and G bands are shown in **Figure 11**.

For an excitation wavelength of 780 nm, the characteristic size L_a, for this sample calculates to a value of 0.17 μm. This characteristic sheet size corresponds with the dimensions for the aggregate samples shown in **Figure 10**. This value is also closer in magnitude to the calculated L_a value for graphite, than that for graphene, due to the greater chance for multiple stacked sheets among the graphene nano-platelets to be responsive to the measurements.

4. Conclusion

To recap, in this chapter we have discussed the ability to discern whether certain graphitic nanomaterials are primarily 2 or 3 dimensional in character, based on fea-tures of their Raman bands. For all three materials (namely graphene, graphite, and functionalized graphene nanoplatelets), we have made use of Tuinstra and Koenig's relationship between the intensities of the D and G Raman bands to characterize the nanomaterials. In addition to the analysis based on Raman spectroscopy, SEM visualization/dimensional analysis was also performed on the graphene nanoplate-let samples. To conclude, the bulk macroscopic 3D character of graphite was clearly apparent compared to the 2D nature of graphene. However, based on the results for the graphene nanoplatelets, both 2D and 3D characteristics/behaviors were present for them, without one dimension dominating the other.

Acknowledgements

Financial support from the National Science Foundation (Award# PHY-1358727 and PHY-1659224) is gratefully acknowledged.

Conflict of interest

No conflict of interest.

Author details

Daniel Casimir, Hawazin Alghamdi, Iman Y. Ahmed, Raul Garcia-Sanchez and Prabhakar Misra*
Laser Spectroscopy Laboratory, Howard University, Washington, DC, USA

*Address all correspondence to: pmisra@howard.edu

IntechOpen

References

[1] Nobelprize.org Physics Prize (1985) [Internet]. 2018. Available from: https://www.nobelprize.org/prizes/physics/1985/summary

[2] Nobelprize.org Physics Prize (1998) [Internet]. 2018. Available from: https://www.nobelprize.org/prizes/physics/1998/summary

[3] Nobelprize.org Physics Prize (2010) [Internet]. 2018. Available from: https://www.nobelprize.org/prizes/physics/2010/press-release

[4] Wong HSP, Akinwande D. Carbon Nanotube and Graphene Device Physics. New York: Cambridge University Press; 2011

[5] Jorio A, Saito R, Dresselhaus G, Dresselhaus M. Raman Spectroscopy in Graphene Related Systems. Weinheim: Wiley-VCH Verlag GmbH & Co. KGaA; 2011

[6] STREM Graphene Nanoplatelets Data Sheet [Internet]. Available from: https://secure.strem.com/uploads/resources/documents/graphene_nanoplatelets_copy1.pdf

[7] Graphene Supermarket Fluorinated Graphene Nanoplatelets Data [Internet]. Available from: https://graphene-supermarket.com/Fluorinated-Graphene-Nanoplatelets.html

[8] Binning G, Rohrer H, Gerber C, Weibel E. Surface studies by scanning tunneling microscopy. Physical Review Letters. 1982;**49**:57

[9] Chieng WB, Ibrahim NA, Yunus WMZW, Hussein MZ, Then YY, Loo YY. Effects of graphene nanoplatelets and reduced graphene oxide on poly(lactic acid) and plasticized poly (lactic acid): A comparative study. Polymers. 2014;**6**:2232-2246

[10] Cunha E, Ren H, Lin F, Kinloch IA, Sun Q, Fan Z, et al. The chemical functionalization of graphene nanoplatelets through solvent-free reaction. Royal Society of Chemistry Advances. 2018;**8**:33564-33572

[11] Hodkiewicz J. Characterizing carbon materials with Raman spectroscopy. Thermo-Fisher Scientific Application Note. 2010;**51901**:1-5

[12] Tuinstra F, Koenig JL. Raman spectrum of graphite. The Journal of Chemical Physics. 1970;**53**(3):1126-1130

[13] Cancado LG, Takai K, Enoki T, Endo M, Kim YA, Mizuasaki H, et al. Measuring the degree of stacking order in graphite by Raman spectroscopy. Carbon. 2008;**46**:272-275

[14] Wojdyr M. Fityk a general-purpose peak fitting program. Journal of Applied Crystallography. 2010;**43**:1126-1128

Structural, Optical and Electrical Properties of Undoped and Doped ZnO Thin Films

Lourdhu Bruno Chandrasekar, S. Nagarajan,
Marimuthu Karunakaran and T. Daniel Thangadurai

Abstract

ZnO, which has high electrochemical stability, wide band gap energy, large excitonic binding energy, intense near band excitonic emission and is non-toxic, have potential applications in all fields. This chapter reviews the structural, optical and electrical properties of undoped and doped ZnO thin films. The type of doping highly influences the structural properties such as grain size, texture coefficient and unit cell properties. The dopants of transition metal and nonmetals have unique characteristics. Moreover, mono-doping and co-doping encourage this research. The optical properties such as bandgap, charge carrier concentrations and transmissions of the films depend on the doping as well as the preparation condition of the films. The effect of doping on its properties is also discussed.

Keywords: structural, electrical, ZnO, doping

1. Introduction

Recent developments in low-dimensional semiconductors create a wide range of applications and opportunities for the fabrication of varieties of devices for the future. The confinement of the charge carrier results in providing unique properties to the materials which are size dependent. Transparent conducting oxides such as CuO, Sn_2O and ZnO have potential applications in the field of photovoltaic and sensing systems. Out of many transparent conducting oxides, ZnO has unique properties such as high excitation binding energy at room temperature, chemical and thermal stability, n-type semiconductor, biocompatibility, etc. [1, 2]. They have a wide range of applications including antibacterial activity, photodetectors, memory applications, LED, solar cell, gas sensors and acoustic wave devices [3–11]. The interests in the ZnO thin films can be viewed easily by the millions of research articles on ZnO thin films prepared under different conditions like substrate temperature, pH value of the solution, doping concentration and annealing temperature. Spray pyrolysis, successive ionic layer adsorption and reaction (SILAR), magnetron sputtering, sol-gel spin coating, pulsed laser deposition, atomic layer deposition and chemical bath deposition are some of the few methods available to prepare the undoped and doped ZnO films [10, 12–17]. Recently, co-doped ZnO such as Al-In-doped ZnO films and F-In-doped ZnO films, ZnO-based thin film transistors, field emission display, white light emission and p-type ZnO encourage

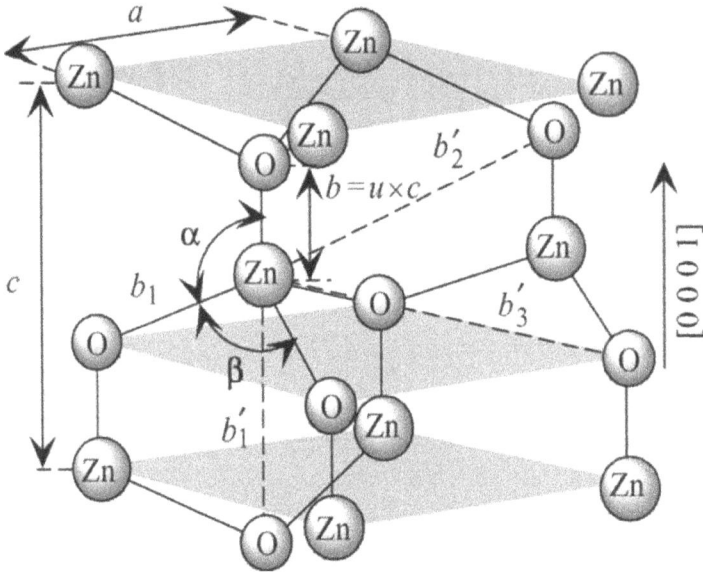

Figure 1.
Representation of wurtzite geometry ZnO unit cell.

this research [18–22]. This chapter provides an insight into the structural, optical and electrical properties of undoped and doped wurtzite geometry ZnO thin films which are quite important for future potential device designs. **Figure 1** shows the representation of wurtzite geometry ZnO. The lattice constants 'a' and 'c' are related by the d-spacing of the corresponding Miller indices (hkl) by $d^{-2} = \frac{4}{3}\left(\frac{h^2+hk+k^2}{a^2}\right) + \frac{l^2}{c^2}$. The ideal value of lattice constants 'a' and 'c' are, respectively, 0.325 and 0.521 nm. The bond length is calculated as $b = \sqrt{\frac{1}{3}a^2 + (0.5 - u)^2 c^2}$, where u is 0.25 + $\{0.33a^2/c^2\}$. The bond angles are $\alpha = \frac{\pi}{2} + arccos\left[\left(\sqrt{1 + 3(c/a)^2 (0.5 - u)^2}\right)^{-1}\right]$ and $\beta = 2\ arcsin\left[\left(\sqrt{\frac{4}{3} + 4(c/a)^2 (0.5 - u)^2}\right)^{-1}\right]$ [23, 24]. It has 3.37 eV band gap at room temperature and possesses both paramagnetic and ferromagnetic nature according to the type of doping.

The structural properties of the ZnO thin films are seldom affected during doping up to certain atomic concentrations, and the structure gets modified slightly due to doping. The structural properties of the doped ZnO thin films are given in Section 2. The changes in optical properties, mainly the band gap variation, are discussed in Section 3. The variations in electrical properties of the undoped and doped thin films are presented in Section 4 followed by the conclusion.

2. Structural properties

Co-doped ZnO thin film is used in transparent electrode applications, where indium-tin-oxide was used in the past. For example, spray pyrolysis is employed to prepare the Co-doped ZnO thin film, and the structural properties are analysed with different doping concentrations [1]. The XRD studies reveal that the wurtzite nature of the materials is retained. The Miller indices (100), (002) and (101) became dominant than (102), (110) and (103) which are dominant for the pristine system. The doping concentration influences the relative intensity observed between the planes (100), (002) and (101). The intensity of (100) peak is maximum when the doping concentration is 0%, and the intensity of (101) peak enhances than (100) peak when the doping concentration is 4%. The variation has been speculated. Nonsystematic variation of grain size is observed as a function of doping concentration. This is because of the fact that the grain size not only depends on the doping concentration but also depends on the crystallographic axes. The packing fraction decreases from 1.612 to 1.602 as the doping concentration increases from 0 to 4% (atomic doping concentration). **Figure 2** shows the grain size and the packing fraction of Co-doped ZnO thin film as a function of Co doping concentration.

Singh et al. reported the Fe- and Ge-doped ZnO thin films prepared by pulsed laser deposition method [25]. The undoped and doped ZnO ($Zn_{0.98}Ga_{0.02}O$ and $Zn_{0.98}Fe_{0.02}O$) thin films show a strong reflection from (002) plane. The grain size corresponds to this miller plane and increases when Ga is doped, and the same decreases when Fe is doped, compared with the undoped ZnO. Moreover, Ga doping increases the texture of the material but decreases the c-lattice parameter. But this situation is reversed due to the doping of Fe. That is, the doping of Fe in ZnO deteriorates the crystalline nature of the material and the c-lattice parameter increases. The same result is obtained when yttrium (Y) is doped in ZnO thin film prepared by sol-gel technique [26]. The reason is due to the fact that ionic radii of Fe^{2+} and Y^{3+} are greater than the ionic radii of Zn^{2+}. In both cases, the addition of either Fe or Y ion shifts the (002) miller plane to the lower angle side. The doping of Y^{3+} in Zn^{2+} lattice increases the c-lattice parameter from 5.205 to 5.247 Å linearly as the function of doping concentration.

The effect of Ni doping in ZnO thin film is reported by Yilmaz [27]. All the films have c-axis orientation which is confirmed by the (002) peaks. The intensity of (002) peak decreases when the doping concentration of Ni increases. The lattice constants are nearly equal irrespective of Ni doping. The full-width at half-maximum (FWHM) increases from 0.266 to 0.344° which indicates that the grain size decreases as the doping of Ni increases. **Figure 3** shows the Ni doping concentration vs. the full-width at half-maximum of the maximum intensity peak. **Figures 4** and **5** indicate the grain size and the dislocation density of the prepared films as a function of doping. The results clearly indicate that the microstructural properties of Ni-doped ZnO thin films are highly influenced by the doping concentration.

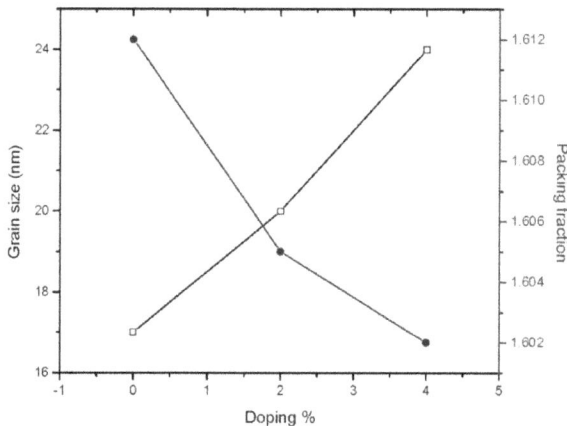

Figure 2.
'Co' doping concentration vs. grain size and packing fraction.

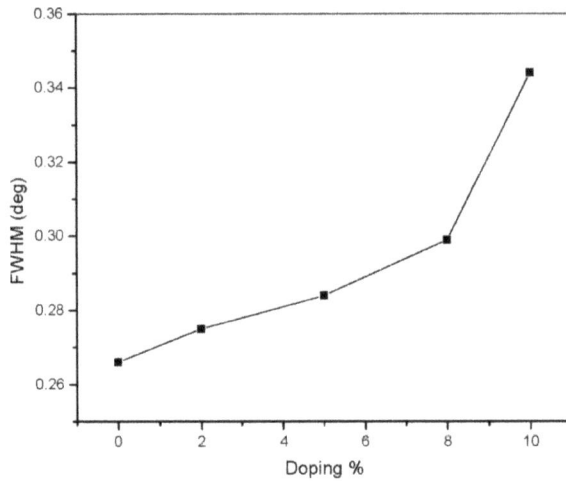

Figure 3.
'Ni' doping concentration vs. FWHM.

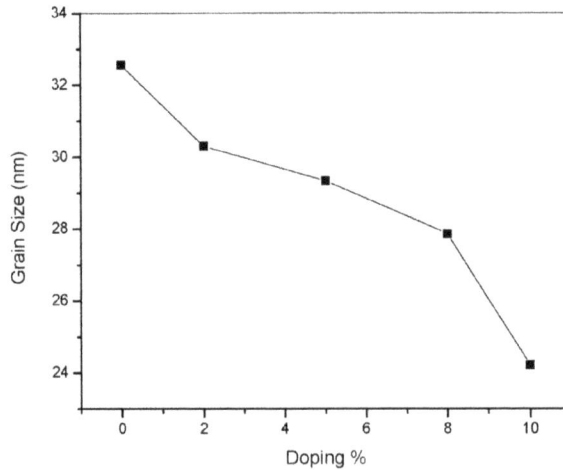

Figure 4.
'Ni' doping concentration vs. grain size.

Ni-doped ZnO thin films are prepared by a modified SILAR method and are reported by Karunakaran et al. [28]. The microstructural properties are analysed as a function of the concentration of nickel sulphate (source material for nickel) and the annealing temperature of the films. As the nickel sulphate concentration changes from 5 to 10 mM, the grain size increases from 29.5 to 44.6 nm. The miller plane (002) has high intensity when the concentration of nickel sulphate is 5 and 10 mM. However, the dominant peak orientation shifts from (002) plane to (101) plane with concentration 15 mM. The annealing increases the crystalline size of the films. The as-deposited films have the average grain size of 45.2 nm, whereas the average grain size is 48.2, 49.7 and 52.8 nm when the films are annealed at 200, 300 and 400°C for 60 min in air, respectively.

Al-doped ZnO thin films, which are prepared by SILAR method, show a redshift towards a higher diffraction angle in the XRD pattern [29]. The line width of the

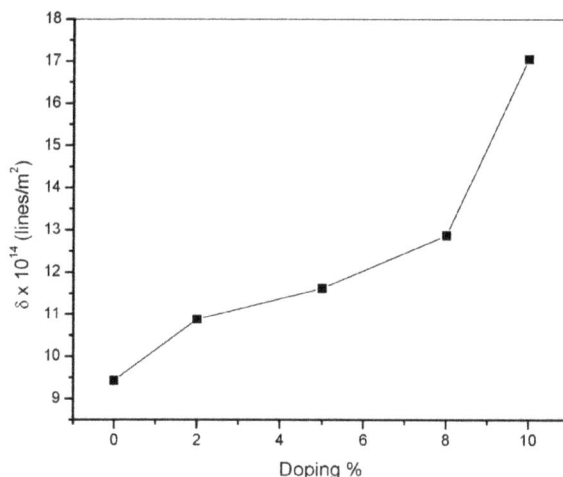

Figure 5.
'Ni' doping concentration vs. dislocation density.

preferred orientation for the undoped ZnO film is 0.18°, and it is 0.21° for Al-doped ZnO films. It is clear evidence that the average grain size decreases from 46 to 39 nm due to the doping of Al. This is due to the contraction of the unit cell and Zn ions which have radius 0.74 Å is replaced by the Al ions (radius 0.51 Å). The absence of the peak (103) in the undoped sample becomes considered as the doping concentration increases. This kind of results is reported in Sr-doped ZnO thin films [16]. The grain size changes from 82 to ~28 nm when the ZnO is doped with Sr. In the Al-doped ZnO films, which are prepared by sol-gel dip coating method, the peak (002) is dominant and becomes more dominant as the concentration of Al increases, but the intensity of the peak deteriorates when 'Al' concentration becomes 10% [30].

The position of the peak (002) shifts to a lower angle side as the doping con-centration increases, and it is shown in **Figure 6**. The peak (103) is dominant in the

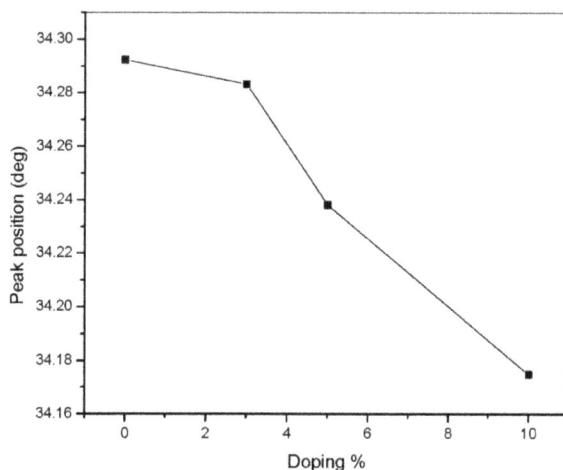

Figure 6.
'Al' doping concentration vs. peak position.

'Al' concentration	Crystallite size from topographical analysis (nm)	Crystallite size corresponds to (002) from XRD (nm)
0	26.0	20.9
3	27.9	16.4
5	33.7	14.0
10	31.5	16.4

Table 1.
Crystallites size of 'Al'-doped ZnO films.

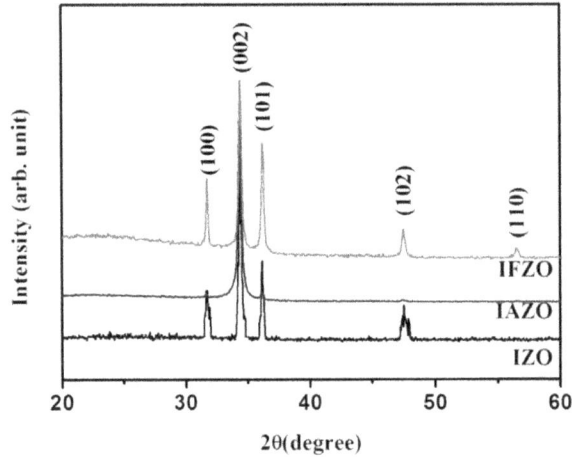

Figure 7.
XRD pattern of IFZO, IAZO and IZO [18].

undoped films and has a negligible intensity as the concentration of 'Al' increases. The crystallite size of the films varies irregularly as the doping of 'Al' concentration changes, and the crystallite sizes are given in **Table 1**.

Recently, co-doped ZnO (In-Al-co-doped ZnO and In-F-doped ZnO) films are prepared by spray pyrolysis technique. **Figure 6** shows the XRD pattern of In-Al-co-doped ZnO (IAZO) and In-F-doped ZnO (IFZO) as compared with In-doped ZnO (IZO) [18].

IFZO films showed a high intensity of (002) diffraction peak compared with other peaks. In addition, IAZO films show only the orientation from (002) peak, which indicates the IAZO films have high stoichiometry than IFZO films. The full-width at half-maximum of the IAZO films corresponding to the miller plane (002) is higher than IFZO films indicating that the IAZO has very low grain size than IFZO. This shows that not only the physical method but also chemical synthesis methods tailoring the structural properties of ZnO thin films with ease by altering parameters or by doping is feasible (**Figure 7**).

3. Optical properties

ZnO films were grown on p^+-Si substrates prepared by atomic layer deposition (ALD) [17]. The photoluminescence spectrum shows the peaks at 376, 426 and 500 nm when it is excited at 250 nm. The peaks at 376, 426 and 500 nm correspond to the band gap 3.3 eV, 'band edge' excitonic states and reflections of the second-order harmonic of the excitation source, respectively. Apart from the second-order harmonic peak, the peak 'band edge' excitonic state is dominant than the peak at 376 nm. When the same film is excited at 390 nm, the peak due to 'band edge' excitonic states is observed at 418 nm. The peak intensities depend on the grown temperature of the material when excited at 250 nm, and the peak intensities remain almost constant when excited at 390 nm.

The Co-doped ZnO thin films show the increase in transmission with a high slope near the fundamental absorption edge in the region of 380–525 nm [1]. The films show high transmittance near the infra-red region. The doping of Co results in the d–d* intrionic transition in the region from 530 to 692 nm. Moreover, the transmission is 63% when the Co doping concentration is 0 and 79% when the concentration is 4%. The direct and indirect band gap of the Co-doped ZnO thin film is shown in **Figure 8**. The band gap decreases as the doping concentration of Co increases.

Siagian et al reported the variation of bandgap in Co-doped ZnO films up to the concentration of 10% [2]. The band gap decreases from 3.337 to 3.099 eV as the doping concentration changes from 0 to 10%, and it is shown in **Figure 9**. The transmittance of the films increases from 68.5 to 83.5% and then decreases as the doping concentration increases. The maximum transmission is obtained for the doping concentration of 7%. Surface roughness, oxygen deficiency and impurity centres are the reasons for a decrease in transmittance when the doping concentration of Co is greater than 7%.

More than 90% of transmittance in the visible region is obtained when Fe and Ga is doped in ZnO thin films [25]. As compared with the undoped ZnO films, Fe doping doesn't change the band gap significantly, but Ga doping increases the band gap. The effect of Ni doping in ZnO on its optical properties is reported by Yilmaz [27]. All the films have a sharp absorption edge at 370 nm. The films show high transparency in the region 550 nm and above. The band gap of the materials is found using dT/dλ and Tauc's plot. The transmission spectra and dT/dλ as a function of wavelength is shown in **Figure 10**. The bandgap of the material decreases as the doping of Ni increases. The transmission and band gap for various Ni doping concentration is given in **Table 2**. But according to the work published by

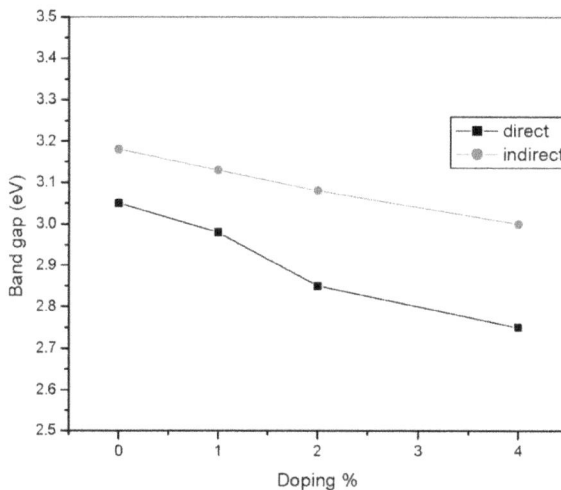

Figure 8.
'Co' doping concentration vs. band gap.

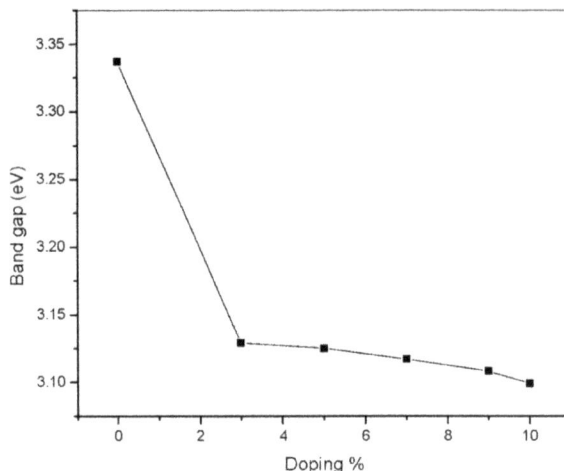

Figure 9.
'Co' doping concentration vs. band gap.

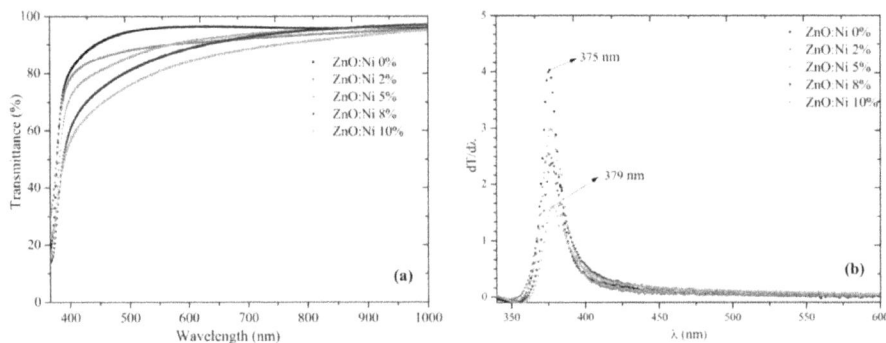

Figure 10.
(a) Transmittance spectra and (b) plot of dT/dλ as a function of wavelength [27].

Ni doping %	Transmittance at 550 nm (%)	Band gap	
		dT/dλ (eV)	Taue's plot (eV)
0	90	3.30	3.27
2	89	3.30	3.26
5	88	3.29	3.23
8	85	3.28	3.22
10	80	3.27	3.19

Table 2.
Transmittance and band gap of Ni-doped ZnO films.

Karunakaran et al., the addition of Ni in Zn lattice increases the band gap [28]. The addition of Ni increases the refractive index and the extinction coefficient.

Y-doped ZnO films show no deep-level emission but show a strong near band edge emission which is due to the recombination of free excitons in the photoluminescence spectrum [26]. The films have a blue shift to higher energy as compared

	ZnO	ZnO:Sr (0.1 mM)	ZnO:Sr (1 mM)
Carrier concentration	$10.01 \times 10^{21}/cm^3$	$27.12 \times 10^{21}/cm^3$	$153.98 \times 10^{21}/cm^3$
Refractive index	2.34	2.33	2.32

Table 3.
Carrier concentration and refractive index of Sr-doped films compared with undoped ZnO films.

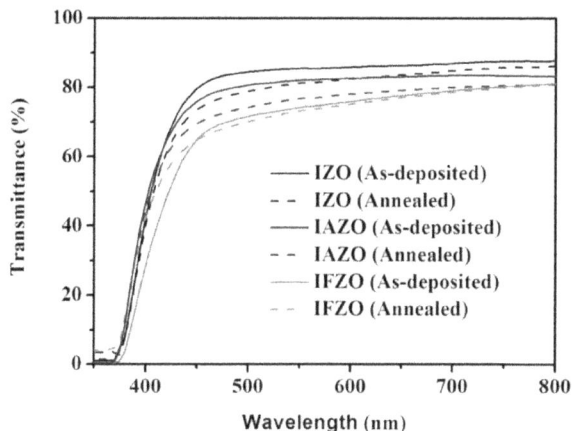

Figure 11.
Transmittance spectra [18].

with undoped ZnO. The near band emission is observed at 3.22 eV when Y concentration is 0% and 3.30 eV when Y is 5%. The Y creates a screened Coulomb potential field, and the results decrease in the intensity of the near band emission.

The band gap increases with an increase in Al concentration in ZnO films [29]. It leads to shifting of absorption band to UV region due to Burstein-Moss shift. The transmittance of doped and undoped ZnO films increase with wavelength. It has a transmittance of 74% in the blue region and more than 90% in the IR region. The decrease in optical transmittance for Al-doped ZnO, when compared with ZnO, is due to the reduction of grain boundaries. The refractive index and extinction coefficient also decreases due to the doping of 'Al'. The transmission of Sr-doped ZnO thin film is higher than undoped ZnO films [16]. Increasing the molarity of the dopant solution increases the transmission coefficient. The optical band gap and the carrier concentration increases, but the refractive index decreases, and the results are given in **Table 3**.

IZO, IAZO and IFZO films show high transmittance between 70 and 85% in the visible region. The annealing process decreases the transmission for IZO and IAZO films, whereas the transmission increases due to the annealing in the case of IFZO thin films. Moreover, the annealing results in a slight blue shift of the absorption edge. **Figure 11** shows the transmission spectra of as-deposited and annealed IZO, IAZO and IFZO films.

4. Electrical properties

ZnO/Si heterojunction diodes show the maximum rectification for 80°C grown ZnO, which has the highest rectification [17]. Such junction has fast ON/OFF switching ratio up to 10^3 times/s. When ZnO is prepared at high temperature, the resistivity of the films becomes comparable to the resistivity of the Si substrates. There is no proper variation observed as a function of the preparation temperature of the films. But as the input wavelength of the photon increases, the photoresponsivity increases. **Table 4** gives the photoresponsivity of the ZnO/Si heterojunction for various grown temperatures and various wavelengths of the incident photon at 0.5 V reverse bias condition.

The Co-doped ZnO thin films show the semiconducting nature, and it is confirmed from the relationship between the conductivity and temperature [1]. The conductivity increases as the temperature increases. The conductivity is measured from 340 K, not from room temperature. The non-linear behaviour of the electrical conductivity is due to the lattice defects. The activation energy decreases due to the increase in the donor carrier density as the doping concentration of Co increases in the temperature limit of 363–403 K. But the activation energy increases due to decrease in Fermi level as the doping concentration increases in the temperature limit 408–473 K. The reported activation energy is given in **Table 5**.

In the case of Y-doped ZnO thin films, the electrical resistivity of the film first decreases and then increases as the doping concentration of Y increases [26]. The minimum electrical resistivity of the films is 7.25 ohm-cm, and this value corre-sponds to the doping concentration of 0.5%. Due to the scattering from grain boundaries and ionized impurities, the Hall mobility decreases gradually from 15.6 to 6.1 $cm^2 V^{-1} s^{-1}$. The carrier concentration of both IAZO and IFZO films are higher than IZO films [18]. This is due to the substitution of fluorine ions at oxygen ion site and aluminium ions at zinc site resulting in one free electron per site. Hence the conductivity of both films increases than IZO films. IFZO films have the highest mobility, and IAZO films have the lowest mobility (**Figure 12**).

	Grown temperature of the heterostructure			
	80°C	150°C	200°C	250°C
λ = 350 nm	37 mAW^{-1}	30 mAW^{-1}	30 mAW^{-1}	35 mAW^{-1}
λ = 475 nm	74 mAW^{-1}	80 mAW^{-1}	80 mAW^{-1}	74 mAW^{-1}
λ = 585 nm	85 mAW^{-1}	90 mAW^{-1}	90 mAW^{-1}	84 mAW^{-1}

Table 4.
Photoresponsivity of the ZnO/Si heterojunction.

Co doping concentration %	Activation energy	
	363–403 K	403–473 K
0	0.405 eV	0.044 eV
1	0.383 eV	0.077 eV
2	0.271 eV	0.343 eV
4	0.119 eV	0.439 eV

Table 5.
Activation energy of Co-doped ZnO films.

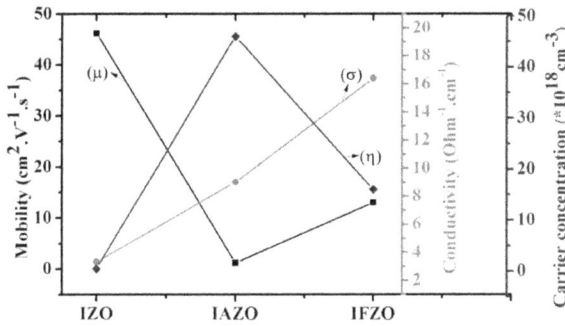

Figure 12.
Mobility, conductivity and carrier concentration of IZO, IAZO and IFZO films [18].

5. Conclusion

This chapter provides an insight into the property tailoring by parameter variations and doping that gives interesting variation in low dimensions for ZnO. Interesting basic properties and wide range of applications encourage the research about undoped and doped ZnO thin film synthesis by both physical and chemical methods. They provide amazing opportunity to include a design with ease to prepare ZnO thin film structures possessing desired electrical and optical properties. Representative works by the author group and few others are reviewed to indicate the variation in structural, optical and electrical properties of undoped and metal-doped ZnO thin films. The co-doped ZnO films are also discussed.

Author details

Lourdhu Bruno Chandrasekar[1*], S. Nagarajan[2], Marimuthu Karunakaran[3] and T. Daniel Thangadurai[4]

1 Department of Physics, The American College, Madurai, India

2 Department of Chemistry, NIT, Imphal, Manipur, India

3 Department of Physics, Alagappa Government Arts College, Karaikudi, India

4 Department of Nanotechnology, Sri Ramakrishna College of Engineering, Coimbatore, India

*Address all correspondence to: brunochandrasekar@gmail.com

IntechOpen

References

[1] Saha SK, Rahman MA, Sarkar MRH, Shahjahan M, Khan MKR. Effect of Co doping on structural, optical, electrical and thermal properties of nanostructured ZnO thin films. Journal of Semiconductors. 2015;**36**:033004

[2] Siagian SM, Sutanto H, Permatasari A. Effect of Co doping to the optical properties of ZnO:Co Thin films deposited on glass substrate by sol-gel spray coating technique. Journal of Physics: Conference Series. 2017;**795**: 012009

[3] Ravichandran AT, Karthick R, Xavier AR, Chandrmohan R, Mantha S. Influence of Sm doped ZnO nanoparticles with enhanced photoluminescence and antibacterial efficiency. Journal of Materials Science: Materials in Electronics. 2017;**28**(9): 6643-6648

[4] Emami-Karvani Z, Chehrazi P. Antibacterial activity of ZnO nanoparticle on Gram-positive and Gram-negative bacteria. African Journal of Microbiology Research. 2011;**5**:1368

[5] Sirelkhatim A, Mahmud S, Seeni A, Mkaus NH, Ann LC, Bakhori SKM, et al. Review on zinc oxide nanoparticles: Antibacterial activity and toxicity mechanism. Nano-Micro Letters. 2015; **7**:219

[6] Hou Y, Mei Z, Du X. Semiconductor ultraviolet photodetectors based on ZnO and MgxZn1-xO. Journal of Physics D: Applied Physics. 2014;**47**:283001

[7] Du XY, Fu YQ, Tan SC, Luo JK, Flewitt AJ, Maeng S, et al. ZnO film for application in surface acoustic wave device. Journal of Physics: Conference Series. 2007;**76**:012035

[8] Zhang H, Shen R, Liang H, Liu Y, Yang Liu X, Xia G. n-ZnO/p-GaN heterojunction light-emitting diodes with a polarization-induced graded-p-AlxGa1−xN electron-blocking layer. Journal of Physics D: Applied Physics. 2013;**46**:065101

[9] Chang S, Park H, Cheng JJ, Rekemeyer PH, Gradecak S. Improved efficiency in organic/inorganic hybrid solar cells by interfacial modification of ZnO nanowires with small molecules. Journal of Physics D: Applied Physics. 2014;**47**:394016

[10] Shishiyanu ST, Shishiyanu TS, Lupan OI. Sensing characteristics of tin-doped ZnO thin films as NO_2 gas sensor. Sensors and Actuators B. 2005;**107**:379

[11] Dighavkar C. Characterization of nanosized zinc oxide based ammonia gas sensor. Archives of Applied Science Research. 2013;**5**:96

[12] Yakuphanoglu F, Ilican S, Caglar M, Caglar Y. The determination of the optical band and optical constants of non-crystalline and crystalline ZnO thin films deposited by spray pyrolysis. Journal of Optoelectronics and Advanced Materials. 2007;**7**:2180

[13] Li W, Hao H. Effect of temperature on the properties of Al:ZnO films deposited by magnetron sputtering with inborn surface texture. Journal of Materials Science. 2012;**47**:3516. DOI: 10.1007/s10853-011-6196-y

[14] Ilican S, Caglar Y, Caglar M. Polycrystalline indium-doped ZnO thin films: Preparation and characterization. Journal of Optoelectronics and Advanced Materials. 2008;**10**:2578

[15] Shukla G, Khare A. Spectroscopic studies of laser ablated ZnO plasma and correlation with pulsed laser deposited ZnO thin film properties. Laser and Particle Beams. 2010;**28**:149

[16] Vijayan TA, Chandramohan R, Valanarasu S, Thirumalai J, Subramanian SP. Comparative investigation on nanocrystal structure, optical, and electrical properties of ZnO and Sr-doped ZnO thin films using chemical bath deposition method. Journal of Materials Science. 2008; **43**:1776

[17] Alkis S, Tekcan B, Nayfeh A, Okyay AK. UV/Vis range photodetectors based on thin film ALD grown ZnO/Si heterojunction diodes. Journal of Optics. 2013;**15**:105002

[18] Hadri A, Taibi M, El Hat A, Mzerd A. Transparent and conductive Al/F and In co-doped ZnO thin films deposited by spray pyrolysis. Journal of Physics: Conference Series. 2016;**689**: 012024

[19] Wei Y, Wei R, Peng S, Zhuangde J. Effect of annealing temperature of $Bi_{1.5}Zn_{1.0}Nb_{1.5}O_7$ gate insulator on performance of ZnO based thin film transistors. Journal of Semiconductors. 2016;**37**:074007

[20] Zulkifli Z, Sharma S, Shinde S, Kalita G, Tannemura M. Effect of annealing in hydrogen atmosphere on ZnO films for field emission display. Materials Science and Engineering. 2015;**99**:0012030

[21] Lv J, Liu C, Gong W, Zi Z, Chen X, Chen X, et al. Study of near white light emission for ZnO thin films grown on silicon substrates. Semiconductor Science and Technology. 2012;27:115021

[22] Jin YP, Zhang B, Wang JZ, Shi LQ. P-type nitrogen-doped ZnO films prepared by in-situ thermal oxidation of Zn_3N_2 films. Chinese Physics Letters. 2016;**33**(5):058101

[23] Bruno Chandrasekar L, Karunakaran M, Chandramohan R, Daniel Thangadurai T, Vijayalakshmi R. X-ray peak profile analysis of Zn_{1-}

$_yMn_yO$ and $Zn_{1-y}Ni_yO$ nanostructures. Journal of Nanoengineering and Nanomanufacturing. 2016;**6**:217

[24] Bruno Chandrasekar L, Chandramohan R, Chandrasekaran S, Thirumalai J, Vijayalakshmi R. Luminescence and unit cell analysis of $Zn_{1-x}Cd_xO$ nanoparticles. Advanced Science Focus. 2013;**1**:292

[25] Singh K, Shukla DK, Majid S, Dhar R, Choudhary RJ, Phase DM. Structural optical and electronic properties of Fe and Ga doped ZnO thin films grown using pulsed laser deposition technique. Journal of Physics: Conference Series. 2016;**755**:012040

[26] Yu Q, Fu W, Yu C, Yang H, Wei R, Sui Y, et al. Ferromagnetic nanocrystalline Gd-doped ZnO powder synthesized by coprecipitation. Journal of Physics D: Applied Physics. 2007; **40**:5592

[27] Yilmaz M. Characteristic properties of spin coated ZnO thin films: The effect of Ni doping. Physica Scripta. 2014;**89**: 095802

[28] Karunakaran M, Chandramohan R, Balamurali S, Gomathi S, Kabila K, Mahalingam T. Current-voltage characteristics of p-CuO/n-ZnO:Sn solar cell. International Journal of Thin Films Science and Technology. 2014;**3**:61

[29] Chandramohn R, Dhanasekaran V, Ezhilvizhian S, Vijayan TA, Thirumalai J, Peter AJ, et al. Spectral properties of aluminium doped zinc oxide thin films prepared by SILAR method. Journal of Materials Science: Materials in Electronics. 2012;**23**:390

[30] Boukhenoufa N, Mahamdi R, Rechem D. Structural, optical, morphological and electrical properties of undoped and Al-doped ZnO thin films prepared using sol–gel dip coating process. Journal of Semiconductors. 2016;**37**:113001

Chapter 4

Van der Waals and Graphene-Like Layers of Silicon Nitride and Aluminum Nitride

Vladimir G. Mansurov, Yurij G. Galitsyn, Timur V. Malin,
Sergey A. Teys, Konstantin S. Zhuravlev, Ildiko Cora and
Bela Pecz

Abstract

A systematic study of kinetics and thermodynamics of Si (111) surface nitridation under ammonia exposure is presented. The appeared silicon nitride (8 × 8) structure is found to be a metastable phase. Experimental evidences of graphene-like nature of the silicon nitride (8 × 8) structure are presented. Interlayer spacings in the $(SiN)_2(AlN)_4$ structure on the Si (111) surface are found equal to 3.3 Å in SiN and 2.86 Å in AlN. These interlayer spacings correspond to weak van der Waals interaction between layers. In contrast to the widely accepted model of a surface structure (8 × 8) as monolayer of β-Si_3N_4 on Si (111) surface, we propose a new graphene-like Si_3N_4 (g-Si_3N_3 and/or g-Si_3N_4) model for the (8 × 8) structure. It is revealed that the deposition of Al atoms on top of a highly ordered (8 × 8) structure results in graphene-like AlN (g-AlN) layers formation. The g-AlN lattice constant of 3.08 Å is found in a good agreement with the *ab initio* calculations. A transformation of the g-AlN to the bulk-like wurtzite AlN is analyzed.

Keywords: kinetics and thermodynamics of 2D layers formation, van der Waals interaction, graphene-like silicon nitride, g-AlN, sp^3- and sp^2-hybridization, π-orbitals, RHEED, STM, STS, HRTEM

1. Introduction

After the discovery of graphene, significant effort is spent to create other graphene-like (graphite-like) materials. Among them, much attention was attracted to graphite-like carbon-nitride compounds g-C_3N_3 and g-C_3N_4 [1–10]. These materials consist of covalently bound sp^2-hybridized carbon and nitrogen atoms. Interest in them is caused by theoretical predictions of new mechanical, electronic, magnetic, and photocatalytic properties [1–9]. To date, the compound g-C_3N_4 has been synthesized [10], and it has been demonstrated that the layers of g-C_3N_4 have a bandgap width in the range of 1.6–2.0 eV, which makes it possible to use the semiconductor layer to create electronics and optoelectronics devices, such as field effect transistors, photodetectors, light-emitting diodes, and lasers. Since silicon is in the same group as carbon in the periodic table, then graphene-like Si-N sheets, where C atoms are replaced by Si atoms, are expected to demonstrate the unusual properties.

Dielectric materials that provide insulation of conductive channels are also necessary for the development of electronic devices. Hexagonal-BN (h-BN), one of the 2D dielectric materials [11, 12], attracts grate attention. However, the fabrication of large area h-BN layers is difficult. AlN is another alternative dielectric material that can be grown epitaxially on large areas. It is also predicted [13] that silicene is stable when encapsulating between two thin graphite-like hexagonal AlN layers. This is especially important, since until now silicene growth has been presented only on metal substrates, which makes it unsuitable for electronic devices. In this chapter, the synthesis and properties of the graphene-like materials and van der Waals layers of silicon nitride (g-SiN) and aluminum nitride (g-AlN) are reported.

2. Kinetics and thermodynamics of g-SiN formation on the Si (111) surface

2.1 Formation kinetics of silicon nitride

The Si_3N_4 film formed on the silicon surface as a rule is amorphous [14–18]. However, at the initial stage of this process, the (8 × 8) structure is formed. The structure (8 × 8) was first discovered by van Bommel and Meyer in 1967 [19]. This structure has been actively studied later. "Modifications" of this structure such as (11/8 × 11/8) and (3/8 × 3/8) have been discovered and described. Models explaining the appearance of (8 × 8) structure by forming a layer of crystalline silicon nitride β–Si_3N_4 are dominating in the literature [20–24].

In our experiments, the nitridation of the silicon surface is started by the onset of ammonia flux onto the clean Si (111) substrate heated to temperatures above 750°C [25]. Two different stages of the silicon nitridation were distinguished by reflection high energy electron diffraction (RHEED): the first stage is a fast formation of the (8 × 8) structure and the following stage is a slow formation of amorphous Si_3N_4 phase. The ordered (8 × 8) structure appears within a few seconds under ammonia flux for the all used temperatures. RHEED pattern of (8 × 8) obtained after exposure of the surface during 6 s under ammonia flux F_{NH3} = 10 sccm at a temperature T = 1050°C is shown in **Figure 1a**. The following bright diffraction spots corresponding to the (8 × 8) structure are clearly observed (**Figure 1a**): (0 -3/8), (0 -5/8), (0 -6/8), (0 -11/8), as well as weaker reflections of (0 -1/8), (0 -2/8), and (0 -7/8) along with the fundamental reflexes (0 0), (0 1), and (0 1) of the Si (111) surface. However, the diffraction spots such as (0 ± 4/8) or (±4/8 ± 4/8) related to the fundamental periodicity of the crystalline phase of β-Si_3N_4 were not observed. This experimental fact indicates that the structure (8 × 8) does not correspond to the β-Si_3N_4 phase, in contrast to the dominating interpretation of nature of the structure (8 × 8) [20–24].

Further nitridation at the same conditions (the second stage) results in silicon nitride amorphous phase (a-Si_3N_4) formation that was accompanied by the total disappearance of all diffraction spots in the RHEED pattern within several minutes. The behavior of intensities of the fractional (0 3/8) diffraction spot as a function of time (i.e., kinetic curves) at different substrate temperatures is shown in **Figure 1b**. **Figure 1b** clearly demonstrates the fast rise of the fractional (0 3/8) spot intensity (the first fast stage) and its further decay of the diffraction spot (the second slow stage). The thickness of a-Si_3N_4 in our experiments was about 5–30 Å, depending on the duration of nitridation. These data do not confirm the possibility of epitaxial growth of crystalline β-Si_3N_4 layers, as was supposed in works [22, 23]. We revealed that the diffraction spot intensity decay as function of time is well described by the exponential law $I(t) = I_0(T) \times exp(-k_2(T) \cdot t)$ at all investigated temperatures T,

(a)

(b)

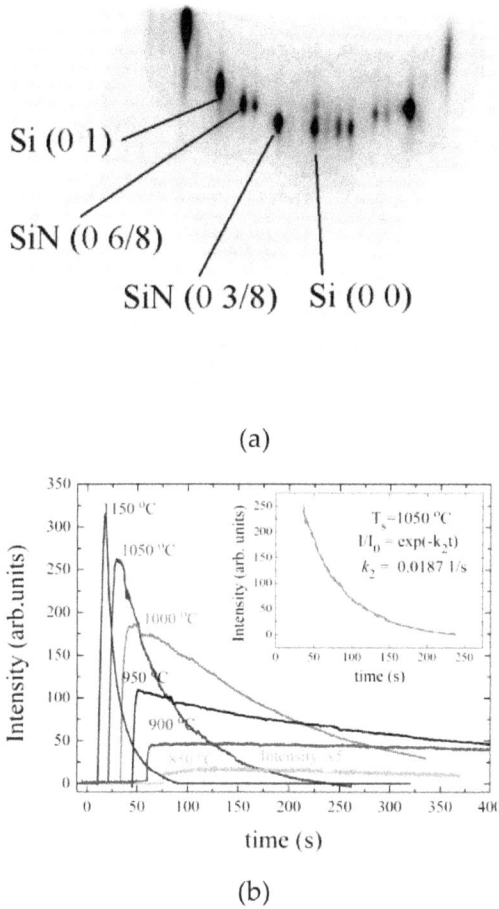

Figure 1.
(a) RHEED pattern of the structure (8 × 8) appeared on the Si (111) substrate after its exposure to 10 sccm ammonia flux for 6 s at temperature of 1050°C; (b) the behavior of fractional (0 3/8) spot intensities at different substrate temperatures, from [25]. The inset **Figure 1b** *shows approximation of the kinetic curve taken at T = 1050°C (gray, solid) by the exponential function (red, dash).*

where k_2 is the rate constant and t denotes time, and so this process corresponds to a first-order reaction. As an example, the inset in **Figure 1b** shows approximation of the experimental curve by the exponential law at T = 1050°C. The activation energy of the amorphous silicon nitride phase formation of 2.4 eV and pre-exponential factor of 10^7–10^8 1/s is found.

Figure 2a shows the normalized kinetic curves for the formation of the (8 × 8) structure, measured by the intensity evolution of the (0 3/8) spot. The figure clearly shows that there is a slight decrease in the rate of formation of the structure (8 × 8) with increasing temperature, which indicates the absence of an activation barrier in this process in contrast to a-Si_3N_4 formation. This fact also does not agree with the formation of a crystalline β-Si_3N_4 layer, which requires the overcoming of a large activation barrier [26].

As shown in work [25], at the formation of the structure (8 × 8), the main role is played by mobile silicon adatoms (Si^a), which are in equilibrium with the surface of the silicon crystal at a given temperature, and the heat of the mobile adatoms formation is 1.7 eV. The existence of mobile adatoms is well known, for example, in the

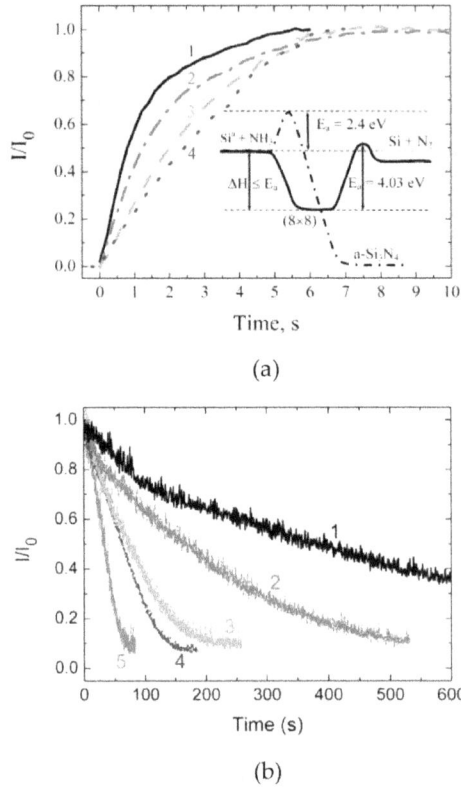

(a)

(b)

Figure 2.
(a) Evolution of the intensity of the (o 3/8) spot during the formation of the (8 × 8) structure at different temperatures: 1. 900°C, 2. 1000°C, 3. 1050°C, 4. 1150°C; (b) kinetic curves of thermal decomposition of the structure (8 × 8): 1. 980°C, 2. 1005°C, 3. 1030°C, 4. 1040°C, 5. 1055°C. All curves are normalized to their own maximum intensity. The inset **Figure 2a** *demonstrates energy diagrams: the solid curve corresponds to formation and decomposition of the structure (8 × 8) and the dashed-dotted curve corresponds to formation of amorphous Si₃N₄.*

temperature range of about 1000°C, and mobile silicon adatoms provide movement of steps on the surface, the formation and/or disappearance of two-dimensional islands, and participate in oxidation processes and other surface reactions [27–29]. Since the rate of formation of the structure (8 × 8) is high during the nitridation process, then the equilibrium concentration of mobile adatoms Si^a does not have time to be established and the coverage of the surface by the two-dimensional phase (8 × 8) is determined by the initial concentration of mobile silicon adatoms at the Si surface at a given temperature. In diffraction, this is manifested in the temperature dependence of the maximum intensity $I_0(T)$. It should be emphasized one more time that the formation of the structure (8 × 8) originates from interaction of ammonia with the mobile silicon adatoms rather than the dangling bonds of silicon atoms incorporated in lattice site (i.e., immobile) on the Si (111) surface.

2.2 Thermal decomposition of a two-dimensional layer (8 × 8)

We investigated the stability of the phase (8 × 8) by studying the kinetics of thermal decomposition of the structure (8 × 8) under ultrahigh vacuum conditions for the temperature range 980–1055°C using the RHEED spots intensity evolution. When the sample was held for several minutes at a fixed temperature, the fractional diffraction spots of the structure (8 × 8) were faded away and the (1 × 1) pattern of the clean silicon surface was restored. **Figure 2b** shows the normalized kinetic curves of the thermal decomposition of the structure (8 × 8) at different temperatures. Analysis of these curves showed that the rate of thermal decomposition increases with increasing temperature, that is, the (8 × 8) structure decomposition is a normal activation process. Curves are well described by a decreasing exponential law $I(t) = I_0 \, exp(-k(T)/t)$, where t is the time, k is the constant of the decomposition rate of the structure (8 × 8), and T is the surface temperature.

The thermal decomposition constant k as function of temperature is presented in the Arrhenius coordinates in **Figure 3**. The activation energy (E_a) of the structure (8 × 8) thermal decomposition, E_a = 4.03 eV, and the pre-exponential factor, k_0 = 2.4 × 10^{13} 1/s, are found. The value of E_a is close to the known binding energy of Si-N bond in Si_3N_4—4.5 eV [30]. Since the activation energy of the decomposition cannot be less than the heat of formation (that is, $E_a \geq \Delta H$), then the heat of formation of the structure (8 × 8) ΔH is no more than 4 eV. The inset of **Figure 2a** schematically illustrates the relationship between the heat of formation and the activation energy of the thermal decomposition of the structure (8 × 8), as well as it shows the energy diagram of the Si_3N_4 amorphous phase formation. The heat of formation of bulk β-Si_3N_4 is about 8 eV [26, 31], which is much larger than the heat of the (8 × 8) structure formation estimated here.

The decomposition rate of the β-Si_3N_4 crystalline phase surface, which was studied in the work [31], is much slower in comparison with the decomposition of the structure (8 × 8), for example, at a temperature of 1740°C, the surface decomposition process took more than an hour. In our case, at a much lower temperature, T = 1055°C, the complete decay of the structure (8 × 8) takes about a minute, which confirms the lower thermal stability of the structure (8 × 8) in comparison with the β-Si_3N_4 crystal. The activation energy of the decomposition of the structure (8 × 8) measured here coincides with the activation energy of the surface thermal decomposition of β-Si_3N_4 (93 kcal/mol), but the pre-exponential factor (2.4 × 10^{13} 1/s) is 10^6 times higher than the pre-exponential factor of β-Si_3N_4 surface decomposition (10^7 1/s) [31]. The coincidence of activation energies shows that in both cases, the limiting stage of the processes is the breaking of the Si-N bonds but the rates of the processes differ by a factor of 10^6. We note that the measured pre-exponential factor has a normal value of ~10^{13} 1/s, which implies a simple decomposition mechanism.

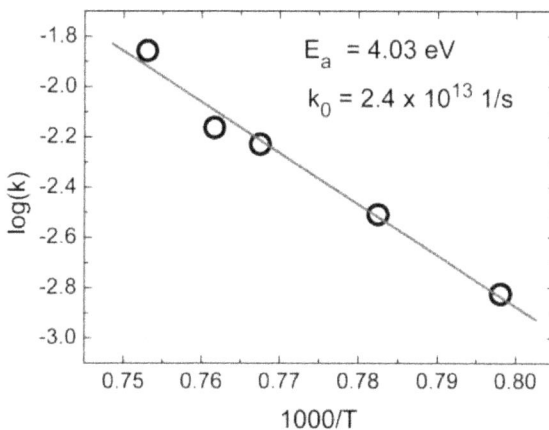

Figure 3.
Arrhenius dependence of the rate constant of the (8 × 8) decomposition.

When the Si-N bonds break, the formation of activated N* nitrogen atoms weakly bound to the surface and following formation of N_2 (N* + N* = N_2) molecules occurs. Thus, the structure (8 × 8) at a temperature above 980°C is destroyed, both during exposure to vacuum and during the continuation of nitridation process under ammonia flux, when it is converted to amorphous Si_3N_4. Indeed, at lower temperatures, the structure (8 × 8) is stable. Therefore, the experimental data on the transformation into an amorphous phase and thermal decomposition evidence the metastability of the phase (8 × 8), in contrast to the stable crystalline phase of β-Si_3N_4 or the amorphous phase of Si_3N_4.

3. Comparative STM/STS study of the (7 × 7) and (8 × 8) structures on the Si (111)

3.1 STM of the Si (111)-(7 × 7) and impact of NH_3 adsorption

The atomic structure of the pristine surface (7 × 7) and the surface with chemisorbed ammonia on Si (111) (obtained at 750°C, 4 min, $P_{NH3} = 10^{-7}$ Torr) were investigated in real space by the scanning tunneling microscopy (STM) method. STM images of these surfaces are presented in **Figure 4a** and **b** (at operating parameters V = +1 V and I = 0.025 nA). The images were obtained in empty electronic states of silicon. Comparing images a and b in **Figure 4**, we can conclude that ammonia is adsorbed mainly to the central adatoms and rest atoms of the structure (7 × 7). In **Figure 4b**, it is also clearly seen that the chemisorption of ammonia induces a disorder on the surface. We consider chemisorption of ammonia as the initial process of nitridation at the silicon surface, followed by the formation of an amorphous nitride phase, since the interaction of ammonia with the dangling bonds of surface silicon atoms (111) does not change the sp^3 hybridization of the orbitals of these atoms. We recall that in the amorphous phase of Si_3N_4, silicon atoms also have sp^3 hybridization of orbitals.

3.2 STS of the Si (111)-(7 × 7) and impact of NH_3 adsorption

We performed measurements of the scanning tunneling spectroscopy (STS) of a clean surface (7 × 7) and on a surface with chemisorbed ammonia (**Figure 4c**). The spectra of pristine silicon surface for various characteristic points, such as corner adatoms, central adatoms, rest atoms, and hole atoms, on the Si (111)-(7 × 7) surface are shown. Each curve for a particular characteristic point is obtained by summing 30–40 volt-ampere curves at equivalent characteristic points on the STM image. One can see a good coincidence of the STS spectra for all these characteristic points on a clean silicon surface. Peaks in the density of states for bias voltages of −0.3, −0.8, −1.5, and −2.3 V, as well as peaks for empty states +0.3 V and +0.8 V, are observed. Similar peaks were observed by many groups [21, 24, 32–37], and they are usually denoted as S_1 = −0.3 eV, S_2 = −0.8 eV, and S_3 = −1.4 eV; we also observed a state at −2.3 eV, which was detected by the XPS method [38]. Some authors associate certain peaks in the density of state spectrum with specific atoms on the surface (7 × 7), for example, peaks at −0.3 and + 0.3 V are associated with adatoms [21, 32], and the peak at −0.8 V is associated with rest atoms [32, 37], that is, they are considered in the framework of the approximation of the local density of electronic states of a given atom. However, our experimental data, namely the presence of identical peaks for the entire family of spectra, for both adatoms and rest atoms and other characteristic points, show that these peaks on the surface of pure silicon should be considered as a manifestation of the surface two-dimensional bands of

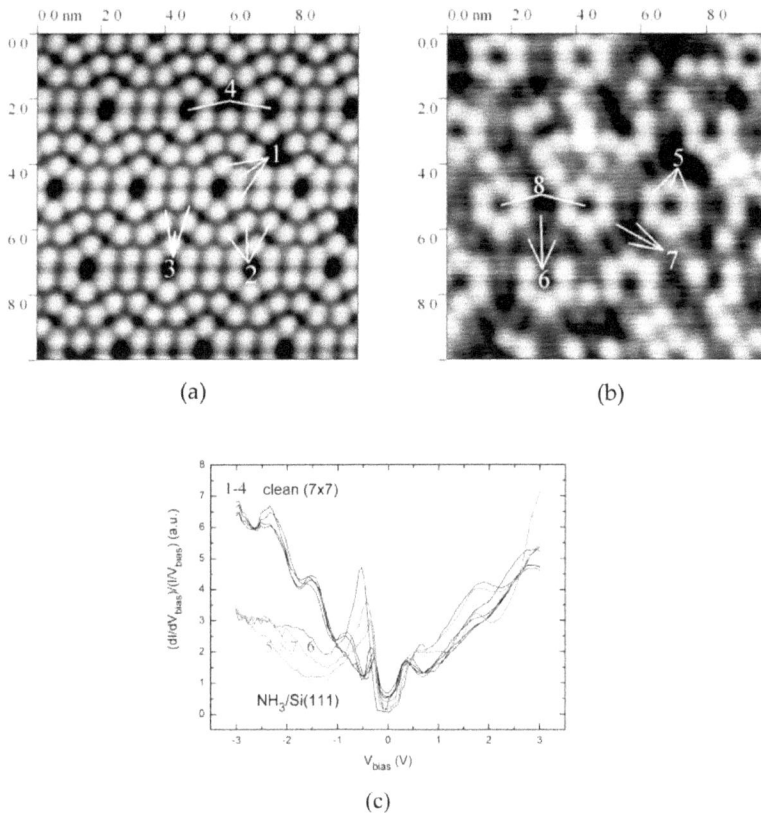

Figure 4.
STM images: (a) the clean surface of Si (111) with reconstruction (7 × 7); (b) the surface of silicon (111) treated with ammonia. The lines indicate the nearest equivalent points. (c) STS spectra of a clean surface of Si (7 × 7) (family of curves 1–4) and spectra after adsorption of ammonia (family of curves 5–8). Curves 1, 5. correspond to corner adatoms; 2, 6. to central adatoms; 3, 7. rest atoms; 4, 8. "corner holes".

states of the structure (7 × 7). Measurements of the STS on the surface with chemisorbed ammonia show different spectra for the same family of the characteristic points. The more pronounced difference in the spectra is observed for corner and central Si adatoms, which was also manifested in STM images, as indicated above. In this case, a stronger local chemical interaction of ammonia with the central adatoms than with the corner adatoms occurs. The surface band structure of the clean surface (7 × 7) is destroyed. Essentially, when random Si-N chemical bonds are formed, various localized electronic states appear.

3.3 STM/STS of the structure (8 × 8)

Interaction of ammonia with the (111) silicon surface at elevated temperatures (800–1150°C) results to (8 × 8) structure, as discussed above, in contrast to disordered structure appeared during adsorption of ammonia at lower temperatures. A typical STM image of the structure (8 × 8) at the working offset $V_s = -3$ V is shown in **Figure 5a**, that is, the image in the filled states of the sample. The periodic structure (8/3 × 8/3) with a distance between the nearest neighboring protrusions $a = 10.2$ Å clearly manifests itself in the figure, in agreement with numerous experimental data, see, for example [22, 23, 39]. In addition, in **Figure 5a**, a honeycomb structure is clearly observed, with a hexagon whose side b is approximately 6 Å in

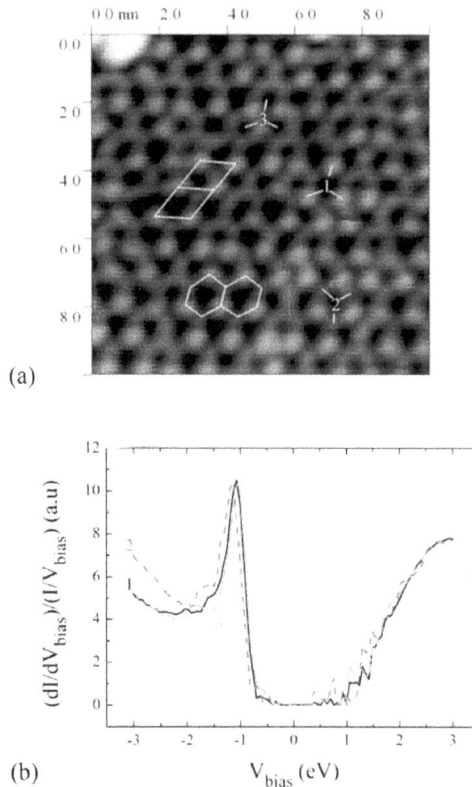

Figure 5.
(a) STM image of the structure (8 × 8). The white figures are marking elementary cells (8/3 × 8/3) and hexagons; (b) STS spectra measured for three characteristic points: 1. protrusion, 2. the vertices of hexagon that not occupied by protrusion, 3. center of hexagon (corresponding three characteristic points are marked in **Figure 5a**).

length and rotated at 30° relative to the unit cell of (8/3 × 8/3). **Figure 5b** clearly shows that the protrusions of the phase (8/3 × 8/3) are brighter than the "vertices" and "sides" of the hexagons. Consequently, the protrusions (8/3 × 8/3) lie on top of the honeycomb structure. In our opinion, protrusions correspond to an ordered adsorption phase, which occupy only three of the six vertices of the hexagons. The relationship between the side length of a hexagon and the distance between protrusions is defined by the expression $a = 2 \times b \times \cos(30°)$. It is clear that the periodicities of adsorption phase and hexagons are the same. There are vacancies in the adsorption phase. This confirms the high mobility of atoms in the adsorption phase, which was noted in the work [24]. At present, it is difficult to unequivocally indicate the nature of the protrusions, perhaps they consist of one, two, or several silicon atoms [40, 41]. For the current study, the hexagonal structure is most interesting, since it determines an atomic arrangement in the structure (8 × 8).

The authors of the work [24] also have observed a honeycomb structure and have explained it by the manifestation of the crystal structure of β-Si_3N_4. The size of the hexagon side in the STM images represented by the authors (see **Figure 5a** and **b** in the article [24]), as well as in our case, was about 6 Å. Let us recall that in the crystal structure of β-Si_3N_4, there is a characteristic fragment—a "small" hexagon with a side of 2.75 Å, as shown experimentally, for example, in work [42] with the help of high-resolution TEM. The hexagonal periodic structure with lattice constant of 7.62 Å of the β-Si_3N_4 is constructed of these "small" hexagons, but there are no hexagons with the 6 Å side in the structure. Consequently, the honeycomb structure detected in this work and in the work [24] does not correspond to the structure of the β-Si_3N_4 crystal. It seems that in most studies devoted to STM of the (8 × 8) structure, only the adsorption phase (8/3 × 8/3) was clearly observed because usually the ordering of the structure is not very high due to the mobility of adatoms (see [21, 24]); hence, the 6 Å honeycomb structure did not be taken into account until now at the modeling of (8 × 8) structure.

Figure 5b shows the STS spectra measured for three different characteristic points: 1. the protrusions, 2. the vertices of the hexagons that not occupied by the protrusions, and 3. the centers of the hexagons. Each curve is obtained by averaging of 30–40 equivalent points. There is a good coincidence of the curves for various characteristic points, as might be expected for a periodic structure. In our opinion, the position of the peak in the density of states at −1.1 eV corresponds to the maximum of the valence band for the surface periodic structure (8 × 8). The band gap of the structure (8 × 8) is about 2.2 eV, and it is determined by the energy gap between the bonding π and antibonding π* orbitals (as discussed below). The band gap of 2.2 eV is much less than the band gap of crystalline β-Si_3N_4 or amorphous Si_3N_4 (4.9–5.3 eV). For comparison, **Figure 6** shows the spectra of a pristine silicon surface with (7 × 7) reconstruction (curve 1), the structure (8 × 8) (curve 2), and a thin amorphous Si_3N_4 layer (curve 3). The STS spectrum of the amorphous phase of Si_3N_4 has a characteristic peak at energy of about −4 eV, which is observed by many groups [21, 24, 43–45]. The authors of [21, 43] refer it to adsorbed nitrogen atoms on the surface, but since this peak exists in thick crystalline β-Si_3N_4 and amorphous Si_3N_4 layers, as demonstrated in works [24, 44, 45], this peak corresponds to the valence-band maximum of bulk Si_3N_4; by the other words, it is the highest occupied molecular orbital (HOMO) of σ bonding band.

The peak at −1.1 eV of the structure (8 × 8) (curve 2) corresponds to the π orbitals, since it has the highest energy among the occupied electron states HOMO and this peak is much higher than peak of σ bonding band (−4 eV) [46]; moreover, this peak is absent in the spectrum of amorphous Si_3N_4. The peak at −1.1 eV was also observed by the method of photoelectron spectroscopy (PES) in [47]. However, the authors attributed it to the dangling silicon or nitrogen bonds of β-Si_3N_4 phase within the framework of the generally accepted concept of the (8 × 8) structure description as a β-Si_3N_4 crystal. However, in **Figure 5** of the work [47],

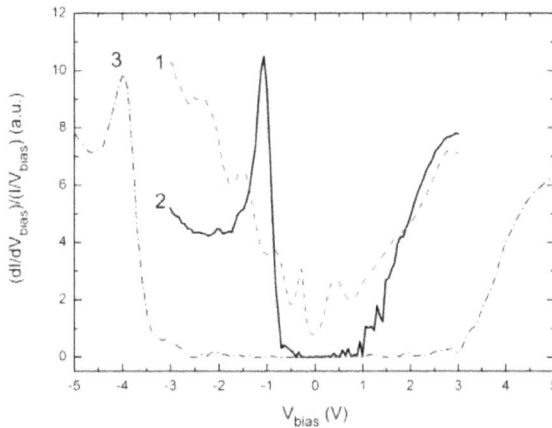

Figure 6.
Scanning tunneling spectra: 1. of the clean silicon surface of Si (111) with reconstruction (7 × 7); 2. of the structure (8 × 8); and 3. of the amorphous phase Si_3N_4.

an appreciable difference in the electronic structures (state densities) β-Si_3N_4 and (8×8) is seen. Moreover, in the works [48, 49], devoted to the calculation of electronic states (0001) β-Si_3N_4, HOMO states associated with dangling bonds did not found. As it will be shown further, it is better to associate this peak in the density of states (8×8) with π-band.

4. HRTEM study of van der Waals structure of silicon nitride and aluminum nitride

The atomic arrangement of (8×8) was also investigated here by the HRTEM method. For these studies, samples with the following sequence of layers were grown on the (111) Si substrate: 2–3 monolayers of silicon nitride with a structure (8×8) and thin epitaxial layer of AlN. The interplanar spacing in the Si substrate in the silicon nitride and AlN layers was determined, see **Figure 7**. It turned out that the interplanar spacing between the layers of silicon nitride and also between the last silicon layer and the silicon nitride layer is about 3.3 Å, which is noticeably larger than the interplanar distances in silicon (3.13 Å) and the known thickness of the β-Si_3N_4 monolayer (2.9 Å). In addition, the layers of silicon nitride differ sharply in contrast from the layers of Si and AlN.

The interplanar distances in silicon nitride of 3.3 Å are larger than the interplanar distances Si 3.13 Å and are larger than the thickness of the monolayer β-Si_3N_4 (2.9 Å). The interplanar distances in the epitaxial AlN layer are also larger than normal interplanar distances in bulk wurtzite AlN (2.49 Å). Therefore, this epitaxial structure $(SiN)_2(AlN)_4$ turned out to be a van der Waals crystal.

Figure 7.
HRTEM image of layers of SiN and AlN on the Si (111) surface.

5. Graphene-like models of the structure (8×8)

The increase in the interplanar distance in silicon nitride, detected by the HRTEM method, is a consequence of the weaker (van der Waals) interaction between the atomic planes. The interaction between the silicon nitride layer and the Si substrate also turned out to be weaker than interaction provided by normal covalent bonds. As mentioned above, when the structure (8×8) is formed, the ammonia interaction occurs with the mobile silicon adatoms, but not with the dangling bonds of the silicon atoms on the surface, which provides an increased interlayer distance between the silicon nitride layer and the silicon surface. Taking into account all the experimental data presented, it can be assumed that the structure (8 × 8) has a graphene-like nature. Moreover, the production of graphene-like AlN layers, described in our paper [50], was possible only on such a graphene-like layer of silicon nitride. If only the silicon nitride layer had dangling bonds (silicon or nitrogen), then the AlN layer would have formed in the bulk wurtzite structure (formation of graphene-like AlN is discussed below). Possible graphene-like models of the layer of silicon nitride g-Si$_3$N$_4$ and g-Si$_3$N$_3$ are shown in **Figure 8**. Similar model g-Si$_3$N$_3$ was considered earlier in the theoretical work of Guo [51]. The basis of both structures is the Si$_3$N$_3$ aromatic conjugated rings connected to each other by either nitrogen atoms (**Figure 8a**) having sp^2 hybridization (as in the β-Si$_3$N$_4$ crystal structure) or via Si-Si bonds (**Figure 8b**). Atoms of silicon in these structures have sp^2 hybridization of atomic orbitals, forming three σ-bonds in planar configuration. The fourth electron of the silicon participates in the π-bond with the nitrogen atom in the ring. Each nitrogen atom uses three valence electrons, and in the aromatic ring, nitrogen has sp. hybridization, and the third valence electron participates in the π-bond.

The proposed structures satisfy the available experimental diffraction data, STM/STS, and HRTEM. They reproduce the periodicity (8 × 8) and the characteristic features of the honeycomb structure observed in the STM, taking into account the weakening of interaction with the silicon surface (there are no dangling bonds in the layer) and explaining the metastability of the structure (8 × 8). Metastability is a consequence of the formation of weaker π-bonds than σ-bonds. In the stable structure of Si$_3$N$_4$ (amorphous or crystalline β-Si$_3$N$_4$), all bonds of silicon and nitrogen are σ-bonds.

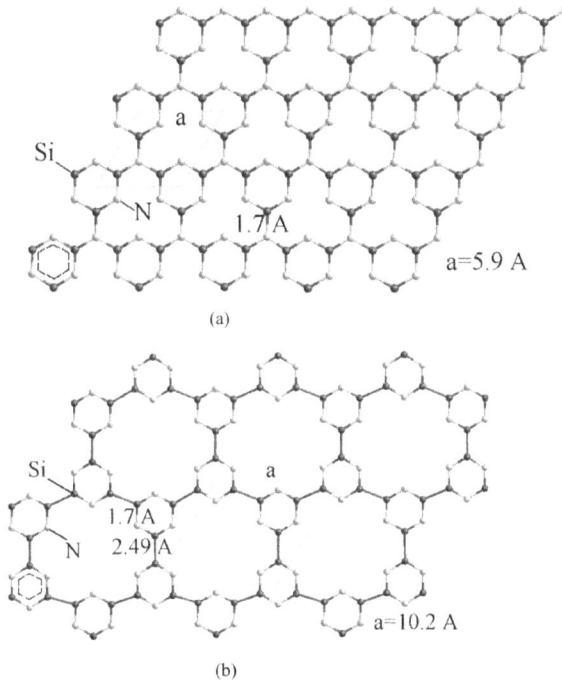

Figure 8.
Models of a graphene-like layer of silicon nitride. (a) g-Si$_3$N$_4$; (b) g-Si$_3$N$_3$.

6. Graphene-like AlN layer formation on Si (111) surface by ammonia MBE

In the present experiments, an AlN flat ultrathin layer was prepared by using the following two-stage procedure. At first, a clean (1 × 1) silicon surface was exposed under the ammonia flux (10 sccm) at the substrate temperatures of 1050°C, and in the second stage, the AlN layer is formed by the Al deposition when the ammonia flux was switched off and the background ammonia pressure of ~10^{-7}–10^{-8} Torr was achieved.

For the AlN formation, the ammonia flux was turned off at the moment when the best (8 × 8) RHEED pattern with sharp and bright eightfold fractional spots was reached. This moment corresponds to the maximum of the curve in **Figure 1b**. Next, the Al deposition onto the highly ordered (8 × 8) structure was performed. The Al flux was established on the value equivalent to the AlN growth rate of ~0.1 Ml/s. The appearance of the AlN diffraction spots and the transformation of the (8 × 8) structure to a new fourfold structure was observed. The RHEED pattern of AlN and (4 × 4) structure is shown in **Figure 9a**. An intensity profile measured along a horizontal line (A-B) crossing the streaks is shown in **Figure 9b**.

The observed fundamental (0-1) AlN streak position exactly coincides with the position of the fractional spot (0-5/4); see **Figure 9b**). Then, an AlN in-plane lattice constant was calculated from the relationship $4 \times a_{111Si} = 5 \times a_{AlN}$, where $a_{111Si} = 3.85$ Å. Hence, the calculated lattice constant is $a_{AlN} = 3.08$ Å. This value

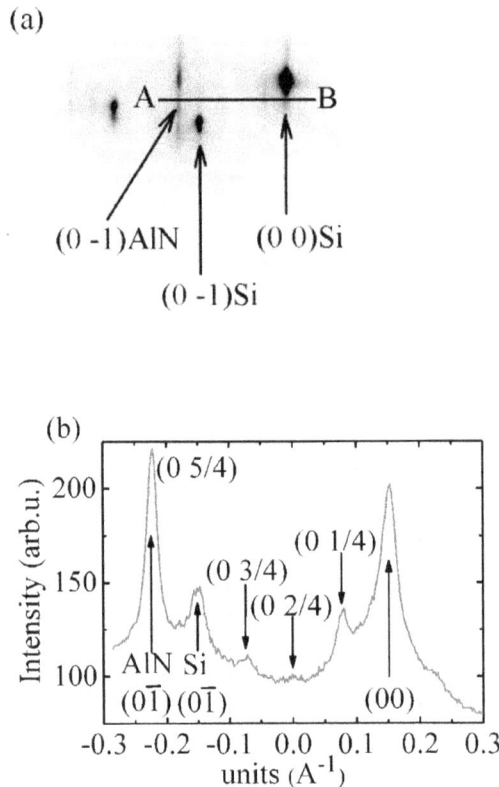

(a)

(b)

Figure 9.
(a) RHEED pattern of the Si surface (4 × 4); (b) the intensity profile of (4 × 4) structure measured along the line (A-B) crossing the diffraction streaks, from [50].

is substantially lower than lateral lattice constant of the bulk wurtzite AlN value being 3.125 Å, which has been measured by the X-ray diffraction method at a high temperature [52].

The (4 × 4) structure is not the consequence of the actual reconstruction of the Si (111) surface. The fractional spots (0 1/4), (0 2/4), and (0 3/4) (and others) are detectable in the RHEED pattern (**Figure 9**) as the result of electrons scattering by both the Si and g-AlN crystal lattices, that is, mixing of reciprocal vectors of these lattices. For example, the fractional reciprocal vectors $q(0\ 1/4)$ is a result of relation $q(0\ 1/4) = q(0\ 1)_{AlN} - q(0\ 1)_{Si}$, where $q(0\ 1)_{AlN}$ and $q(0\ 1)_{Si}$ are integer order reciprocal vectors of g-AlN and Si, respectively. The approximate equality of the fourfold silicon lateral constant and the fivefold wurtzite AlN lateral constant was previously pointed out in some studies (e.g., see paper [53, 54]) with mismatch of ~1.3%. In our case, there is an exact coincidence, and as the result, the fractional beams of (4 × 4) structure are experimentally observed in the RHEED pattern. Thus, an extremely thin and flat of AlN on (111) Si substrate with the lattice constant of a = 3.08 Å is prepared. This value is close to the *ab initio* calculated value of 3.09 Å for the graphene-like aluminum nitride lattice with sp^2-like bonding [55, 56].

There is only one more experimental work [57] that dedicated to epitaxial growth of graphite-like hexagonal AlN nanosheets on single crystal Ag (111). It is interesting to note that a simple our calculation of the g-AlN lattice constant using the lateral constant of (111) Ag (2.89 Å) from the diffraction pattern presented by authors of the work gives the value in the range of 3.06–3.09 Å in contrast to the value presented by the authors of 3.13 Å.

We have carried out experimental precise measurements of the in-plane AlN lattice constant under the Al and NH_3 fluxes supplied onto the surface either separately or simultaneously. The evolution of in-plane lattice constant during the g-AlN formation process is depicted in **Figure 10**. The increasing of the lattice constant from 3.08 to 3.09 Å under Al flux without ammonia flux (i.e., there is no growth of AlN) is clearly visible.

Then, the Al flux was turned off (at the moment of 340 s) and the NH_3 flux (10 sccm) was switched on. The lateral lattice constant of g-AlN was keeping the same value of 3.09 Å, and so, under the ammonia flux, the formed g-AlN is quite stable. The lateral size of the g-AlN islands has been estimated from the RHEED data as ~70–100 Å.

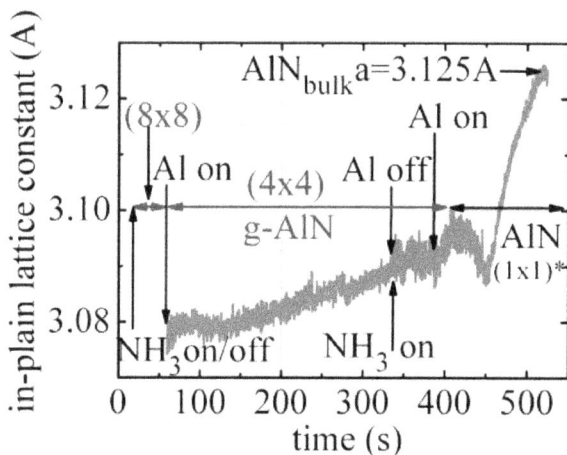

Figure 10.
Evolution of an in-plane lattice constant during the g-AlN formation process.

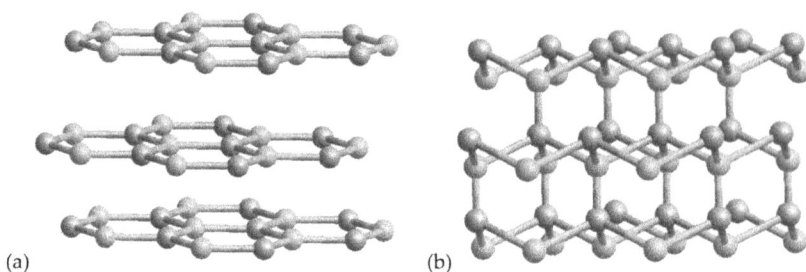

(a)   (b)

Figure 11.
Atomic models of (a) graphene-like AlN; (b) wurtzite AlN.

The epitaxial growth of AlN was initiated by turning on the Al flux, keeping the same ammonia flux (at the moment of 390 s). The fractional streaks of the (4 × 4) structure are gradually dimmed together with the increasing of the fundamental (0 1) g-AlN streak intensity. At the moment of about 420 s, a structure (1 × 1) of g-AlN appears. Further growth of AlN leads to the lattice constant conversion from 3.09 to 3.125 Å. The value 3.125 Å corresponds to bulk value of the wurtzite AlN lattice constant at high temperature [52].

Thus, the transformation from graphite-like (sp^2-hybridization, see **Figure 11a**) to wurtzite structure of AlN (sp^3-hybridization, **Figure 11b**) is observed. This transition is similar to the transition metastable graphene-like silicon nitride of the structure (8 × 8) to stable amorphous phase a-Si$_3$N$_4$. The maximal thickness of the g-AlN layer of ~5–6 AlN monolayers was estimated using the AlN growth rate that is less than the theoretically predicted value of 22–24 monolayer [58]. The difference between calculations and experimental data might be attributed to the competition with the bulk stabilization mechanism involving structural defects and roughening, which were not taken into account in work [58]. The similar discrepancy between theoretically predicted and experimentally measured thickness of the sp^2-sp^3 transition was noticed for ZnO [59].

7. Conclusion

Systematic studies of the structure (8 × 8) by the methods RHEED, STM/STS, and HRTEM were carried out. It is found that the structure (8 × 8) is formed within 5–7 s during nitridation of the Si (111) surface at the temperature range 950–1150°C. The formation rate of the structure (8 × 8) is independent on the temperature. The kinetics of the thermal decomposition of this two-dimensional layer of silicon nitride has been studied. It is established that the structure (8 × 8) is a metastable phase, and with further nitridation, a transition to the stable amorphous phase of Si$_3$N$_4$ occurs. In the structure (8 × 8), the honeycomb structure with the side length of hexagon of 6 Å was found for the first time, on which the adsorption phase of silicon is located with a periodicity of 10.2 A. The interplanar spacing in the epitaxial structure (SiN)$_2$(AlN)$_4$ on the (111) Si surface are measured: 3.3 Å in silicon nitride layer and 2.86 Å in AlN. These interlayer distances correspond to the weak van der Waals interaction between the layers. Scanning tunneling spectroscopy in the filled states revealed a peak of 1.1 eV below the Fermi level. Comparison of the measurements in the STS of the metastable phase of silicon nitride with measurements made on a clean surface of Si (111)–(7 × 7) and on an amorphous Si$_3$N$_4$ layer helped us in identifying the peak −1.1 eV as the π-bonding band of the structure (8 × 8). The band gap between the bonding and antibonding orbitals is

	Lateral lattice constant	Interplanar spacing	Heat of formation	Band gap
g-Si$_3$N$_3$ (g-Si$_3$N$_4$)	10.2 Å (5.9 Å hex side)	3.3 Å	< 4 eV	2.2 eV
g-AlN	3.08 Å	2.86 Å		~ 3.0 eV [55]

Table 1.
Structural and thermodynamic characteristics of g-AlN and g-Si$_3$N$_3$.

found of 2.2 eV. Consequently, the 2D silicon nitride layer is a semiconductor. The data obtained allowed us to propose new graphene-like model structures (8 × 8). The models are planar graphene-like structures of g-Si$_3$N$_3$ and/or g-Si$_3$N$_4$. Owing to the formation of graphene-like Si$_3$N$_3$ layer, it is possible to synthesize graphene-like g-AlN, the lateral lattice constant of 3.08 Å, and the interplanar distance of 2.86 Å. When the AlN thickness of 5–6 monolayers is reached, g-AlN passes into the wurtzite structure.

Table 1 summarizes some important characteristics of graphene-like phases of silicon nitride and aluminum nitride.

Acknowledgements

This work was supported by the Russian Foundation for Basic Research (Grant Nos. 17-02-00947, 18-52-00008).

Author details

Vladimir G. Mansurov[1*], Yurij G. Galitsyn[1], Timur V. Malin[1], Sergey A. Teys[1], Konstantin S. Zhuravlev[1], Ildiko Cora[2] and Bela Pecz[2]

1 Rzhanov Institute of Semiconductor Physics Siberian Branch of Russian Academy of Sciences, Novosibirsk, Russia

2 Thin Film Physics Department, Institute of Technical Physics and Materials Science Centre for Energy Research, Hungarian Academy of Sciences, Budapest, Hungary

*Address all correspondence to: mansurov@isp.nsc.ru

IntechOpen

References

[1] Du A, Sanvito S, Smith SC. First-principles prediction of metal-free magnetism and intrinsic half-metallicity in graphitic carbon nitride. Physical Review Letters. 2012;**108**:197207

[2] Huynh MH, Hiskey MA, Archuleta JG, Roemer EL. Preparation of nitrogen-rich nanolayered, nanoclustered, and nanodendritic carbon nitrides. Angewandte Chemie. 2005;**117**:747-749

[3] Li X, Zhang S, Wang Q. Stability and physical properties of a tri-ring based porous g-C_4N_3 sheet. Physical Chemistry Chemical Physics. 2013;**15**:7142-7146

[4] Niu P, Liu G, Cheng HM. Nitrogen vacancy-promoted photocatalytic activity of graphitic carbon nitride. Journal of Physical Chemistry C. 2012;**116**:11013-11018

[5] Niu P, Zhang L, Liu G, Cheng HM. Graphene-like carbon nitride nanosheets for improved photocatalytic activities. Advanced Functional Materials. 2012;**22**:4763-4770

[6] Wang X, Blechert S, Antonietti M. Polymeric graphitic carbon nitride for heterogeneous photocatalysis. ACS Catalysis. 2012;**2**:1596-1606

[7] Li X, Zhou J, Wang Q, Kawazoe Y, Jena P. Patterning graphitic C−N sheets into a kagome lattice for magnetic materials. Journal of Physical Chemistry Letters. 2013;**4**:259-263

[8] Teter DM, Hemley RJ. Low-compressibility carbon nitrides. Science. 1996;**271**:53-55

[9] Botari T, Huhn WP, Lau VW, Lotsch BV, Blum V. Thermodynamic equilibria in carbon nitride photocatalyst materials and conditions for the existence of graphitic carbon nitride g-C_3N_4. Chemistry of Materials. 2017;**29**:4445-4453

[10] Algara-Siller G, Severin N, Chong SY, Björkman T, Palgrave RG, Laybourn A, et al. Triazine-based graphitic carbon nitride: A two-dimensional semiconductor. Angewandte Chemie. 2014;**53**:7450-7455

[11] Kim KK, Hsu A, Jia X, Kim SM, Shi Y, Hofmann M, et al. Synthesis of monolayer hexagonal boron nitride on Cu foil using chemical vapor deposition. Nano Letters. 2012;**12**:161-166

[12] Ismach A, Chou H, Ferrer DA, Wu Y, McDonnell S, Floresca HC, et al. Toward the controlled synthesis of hexagonal boron nitride films. ACS Nano. 2012;**6**(7):6378-6385

[13] Houssa M, Pourtois G, Afanas'ev VV, Stesmans A. Can silicon behave like graphene? A first-principles study. Applied Physics Letters. 2010;**97**:112106

[14] Moslehi MM, Saraswat KC. Thermal nitridation of Si and SiO_2 for VLSI. IEEE Journal of Solid-State Circuits. 1985;**20**:26-43

[15] Murarka SP, Chang CC, Adams AC. Thermal nitridation of silicon in ammonia gas: Composition and oxidation resistance of the resulting films. Journal of the Electrochemical Society. 1979;**126**:996-1003

[16] Hayafuji Y, Kajiwara K. Nitridation of silicon and oxidized-silicon. Journal of the Electrochemical Society. 1982;**129**:2102-2108

[17] Wu CY, King CW, Lee MK, Tchen CT. Growth kinetics of silicon thermal nitridation. Journal of the Electrochemical Society. 1982;**129**:1559-1563

[18] Schrott AG, Fain SC. Nitridation of Si(111) by nitrogen atoms II. Surface Science. 1982;**123**:204-222

[19] van Bommel AJ, Meyer F. A low energy electron diffraction study of the PH_3 adsorption on the Si (111) surface. Surface Science. 1967;**8**:381-398

[20] Rottger B, Kliese R, Neddermeyer H. Adsorption and reaction of NO on Si(111) studied by scanning tunneling microscopy. Journal of Vacuum Science and Technology B. 1996;**14**:1051-1054

[21] Ahn H, Wu CL, Gwo S, Wei CM, Chou YC. Structure determination of the Si_3N_4/Si(111)-(8×8) Surface: A combined study of kikuchi electron holography, scanning tunneling microscopy, and *ab initio* calculations. Physical Review Letters. 2001;**86**:2818-2821

[22] Wang XS, Zhai G, Yang J, Cue N. Crystalline Si_3N_4 thin films on Si(111) and the 4×4 reconstruction on Si_3N_4(0001). Physical Review B. 1999;**60**:R2146-R2149

[23] Wang XS, Zhai G, Yang J, Wang L, Hu Y, Li Z, et al. Nitridation of Si(111). Surface Science. 2001;**494**:83-94

[24] Flammini R, Allegrini P, Wiame F, Belkhou R, Ronci F, Colonna S, et al. Nearly-free electron like surface resonance of a β-Si_3N_4(0001)/Si(111)-8×8 interface. Physical Review B. 2015;**91**:075303

[25] Mansurov VG, Malin TV, Galitsyn YG, Shklyaev AA, Zhuravlev KS. Kinetics and thermodynamics of Si(111) surface nitridation in ammonia. Journal of Crystal Growth. 2016;**441**:12-17

[26] Jennings HM, Edwards JO, Richman MH. Molecular structure, microstructure, macrostructure and properties of silicon nitride. Inorganica Chimica Acta. 1976;**20**:167-181

[27] Rogilo DI, Fedina LI, Kosolobov SS, Ranguelov BS, Latyshev AV. Critical terrace width for two-dimensional

nucleation during Si growth on Si(111)-(7×7) surface. Physical Review Letters. 2013;**111**:036105

[28] Sitnikov S, Kosolobov S, Latyshev A. Attachment–detachment limited kinetics on ultra-flat Si(111) surface under etching with molecular oxygen at elevated temperatures. Surface Science. 2015;**633**:L1-L5

[29] Sitnikov SV, Latyshev AV, Kosolobov SS. Advacancy-mediated atomic steps kinetics and two-dimensional negative island nucleation on ultra-flat Si(111) surface. Journal of Crystal Growth. 2017;**457**:196-201

[30] Shetty S, Kesaria M, Ghatak J, Shivaprasad SM. The origin of shape, orientation, and structure of Spontaneously formed Wurtzite GaN nanorods on cubic Si(001) Surface. Crystal Growth and Design. 2013;**13**:2407-2412

[31] Batha HD, Whitney ED. Kinetics and mechanism of the thermal decomposition of Si_3N_4. Journal of the American Ceramic Society. 1973;**56**:365-369

[32] Wolkow R, Avouris P. Atom-resolved surface chemistry using scanning tunneling microscopy. Physical Review Letters. 1988;**60**:1049-1052

[33] Brommer KD, Galvan M, Dal Pino A, Joannopoulos JD. Theory of adsorption of atoms and molecules on Si(111)-(7×7). Surface Science. 1994;**314**:57-70

[34] Lim H, Cho K, Park I, Joannopoulos JD, Kaxiras E. *Ab initio* study of hydrogen adsorption on the Si(111)-(7×7) surface. Physical Review B. 1995;**52**:17231-17237

[35] Chen L, Pan BC, Xiang H, Wang B, Yang J, Hou JG, et al. Observation of local electronic structures of adatom

vacancies in Si(111)−(7×7) surface in real space. Physical Review B. 2007;**75**:085329

[36] Odobescu AB, Maizlakh AA, Zaitsev-Zotov SV. Electron correlation effects in transport and tunneling spectroscopy of the Si(111)−7×7 surface. Physical Review B. 2015;**92**:165313

[37] Siriwardena HD, Yamashita T, Shimomura M. STM Observation of the Si(111)-(7×7) reconstructed surface modified by excess phosphorus doping. International Journal of Electrical and Computer Engineering. 2017;**7**:2993-3001

[38] Losio R, Altmann KN, Himpsel FJ. Continuous transition from two- to one-dimensional states in Si(111)-(5×2)−Au. Physical Review Letters. 2000;**85**:808-811

[39] Morita Y, Tokumoto H. Origin of the 8/3×8/3 superstructure in STM images of the Si(111)-8×8: N surface. Surface Science Letters. 1999;**443**:L1037-L1042

[40] Binggeli N, Chelikowsky JR. Langevin molecular dynamics with quantum forces: Application to silicon clusters. Physical Review B. 1994;**50**:11764-11770

[41] Dixon DA, Feller D, Peterson KA, Gole JL. The molecular structure and ionization potential of Si_2: The role of the excited states in the photoionization of Si_2. Journal of Physical Chemistry A. 2000;**104**:2326-2332

[42] Ziegler A, Kisielowski C, Ritchie RO. Imaging of the crystal structure of silicon nitride at 0.8 Angstrom resolution. Acta Materialia. 2002;**50**:565-574

[43] Wu C, Chen W, Su Y. N2-plasma nitridation on Si(111): Its effect on crystalline silicon nitride growth. Surface Science. 2012;**606**:L51-L54

[44] Xu Y, Ching WY. Electronic structure and optical properties of α and β phases of silicon nitride, silicon oxynitride, and with comparison to silicon dioxide. Physical Review B. 1995;**51**:17379-17389

[45] Gritsenko VA. Electronic structure of silicon nitride. Physics-Uspekhi. 2012;**55**:498-507

[46] Giuliani M, Motta N. Polimer self-assembly on carbon nanotubes. In: Belucci S, editor. Self-Assembly of Nanostructures The INFN lactures Vol. III. New York: Springer; 2012

[47] Kim JW, Yeom HW. Surface and interface structures of epitaxial silicon nitride on Si(111). Physical Review B. 2003;**67**:035304

[48] Bagatur'yants AA, Novoselov KP, Safonov AA, Cole JV, Stoker M, Korkin AA. Silicon nitride chemical vapor deposition from dichlorosilane and ammonia: Theoretical study of surface structures and reaction mechanism. Surface Science. 2001;**486**:213-225

[49] Bermudez VM. Theoretical study of the electronic structure of the $Si_3N_4(0001)$ surface. Surface Science. 2005;**579**:11-20

[50] Mansurov V, Malin T, Galitsyn Y, Zhuravlev K. Graphene-like AlN layer formation on (111)Si surface by ammonia molecular beam epitaxy. Journal of Crystal Growth. 2015;**428**:93-97

[51] Guo Y, Zhang S, Wang Q. Electronic and optical properties of silicon based porous sheets. Physical Chemistry Chemical Physics. 2014;**16**:16832-16836. DOI: 10.1039/C4CP01491J

[52] Yim WM, Paff RJ. Thermal expansion of AlN, sapphire, and silicon. Journal of Applied Physics. 1974;**45**:1456-1457

[53] Wu C, Wang J, Chan MH, Chen TT, Gwo S. Heteroepitaxy of GaN on Si(111) realized with a coincident-interface AlN/β-Si₃N₄(0001) double-buffer structure. Applied Physics Letters. 2003;**83**:4530-4532

[54] Bourret A, Barski A, Rouviere JL, Renaud G, Barbier A. Growth of aluminum nitride on (111) silicon: Microstructure and interface structure. Applied Physics Letters. 1998;**83**:2003-2009

[55] Şahin H, Cahangirov S, Topsakal M, Bekaroglu E, Akturk E, Senger RT, et al. Monolayer honeycomb structures of group-IV elements and III-V binary compounds: First-principles calculations. Physical Review B. 2009;**80**:155453

[56] Mukhopadhyay G, Behera H. Structural and Electronic Properties of Graphene and Graphene-like Materials. 2012. Available from: http://arxiv.org/abs/1210.3308

[57] Tsipas P, Kassavetis S, Tsoutsou D, Xenogiannopoulou E, Golias E, Giamini SA, et al. Evidence for graphite-like hexagonal AlN nanosheets epitaxially grown on single crystal Ag(111). Applied Physics Letters. 2013;**103**:251605

[58] Freeman CL, Claeyssens F, Allan NL, Harding JH. Graphitic nanofilms as precursors to Wurtzite films: Theory. Physical Review Letters. 2006;**96**:066102

[59] Tusche C, Meyerheim HL, Kirschner J. Observation of depolarized ZnO (0001) monolayers: Formation of unreconstructed planar sheets. Physical Review Letters. 2007;**99**:026102

Chapter 5

Polarizability and Impurity Screening for Phosphorene

Po Hsin Shih, Thi Nga Do, Godfrey Gumbs and Dipendra Dahal

Abstract

Using a tight-binding Hamiltonian for phosphorene, we have calculated the real part of the polarizability and the corresponding dielectric function, $\text{Re}[\epsilon(\mathbf{q}, \omega)]$, at absolute zero temperature (T = 0 K) with free carrier density $10^{13}/\text{cm}^2$. We present results showing $\text{Re}[\epsilon(\mathbf{q}, \omega)]$ in different directions of the transferred momentum q. When q is larger than a particular value which is twice the Fermi momentum k_F, Re $[\epsilon(\mathbf{q}, \omega)]$ becomes strongly dependent on the direction of $\mathbf{q}$. We also discuss the case at room temperature (T = 300 K). These results which are similar to those previously reported by other authors are then employed to determine the static shielding of an impurity in the vicinity of phosphorene.

Keywords: phosphorene, polarizability, impurity screening

1. Introduction

Emerging phenomena in physics and quantum information technology have relied extensively on the collective properties of low-dimensional materials such as two-dimensional (2D) and few-layer structures with nanoscale thickness. There, the Coulomb and/or atomic interactions play a crucial role in these complexes which include doped as well as undoped graphene [1–3], silicene [4, 5], phosphorene [6, 7], germanene [8, 9], antimonene [10, 11], tinene [12], bismuthene [13–18] and most recently the 2D pseudospin-1 $\alpha - T_3$ lattice [19]. Of these which have been successfully synthesized by various experimental techniques and which have been extensively investigated by various experimental techniques, few-layer black phosphorus (phosphorene) or BP has been produced by using mechanical cleavage [6, 20], liquid exfoliation [7, 21, 22], and mineralizer-assisted short-way transport reaction [23–25].

Unlike graphene, phosphorus inherently has an appreciable band gap. The observed photoluminescence peak of single-layer phosphorus in the visible optical range shows that its band gap is larger than that for bulk. Furthermore, BP has a middle energy gap (~1.5–2 eV) at the Γ point, thereby being quite different from the narrow or zero gaps of group-IV systems. Specifically, experimental measurements have shown that the BP-based field effect transistor has an on/off ratio of 105 and a carrier mobility at room temperature as large as 103 cm^2/Vs. We note that BP is expected to play an important role in the next-generation of electronic devices [6, 20]. Phosphorene exhibits a puckered structure related to the sp^3 hybridization of ($3s$, $3p_x$, $3p_y$, $3p_z$) orbitals. The deformed hexagonal lattice of monolayer BP has four atoms [26], while the group-IV honeycomb lattice includes two atoms. The low-lying energy dispersions, which are dominated by $3p_z$ orbitals, can be described by a four-band model with complicated multi-hopping integrals [26]. The low-lying energy bands are highly anisotropic, e.g., the linear and parabolic dispersions near the Fermi energy E_F, respectively, along the $\hat{k}_x$ and $\hat{k}_y$ directions. The anisotropic behaviors are further reflected in other physical properties, as verified by recent measurements on optical and excitonic spectra [27] as well as transport properties [6, 28].

In this work, we have examined the anisotropic behavior of the static polarizability and shielded potential of an impurity for BP. The calculations for the polarizability were executed at T = 0 K and room temperature (T = 300 K). We treat the buckled BP structure as a 2D sheet in our formalism. Consequently, we present an algebraic expression for the surface response function of a pair of 2D layers with arbitrary separation and which are embedded in dielectric media. We then adapt this result to the case when the layer separation is very small to model a free-standing buckled BP structure.

• The outline of the rest of our presentation is as follows. In Section 2, we present the surface response function for a pair of 2D layers embedded in background dielectric media. We then simplify this result for a pair of planar sheets which are infinitesimally close to each other and use this for buckled BP. The tight-binding model Hamiltonian for BP is presented in Section 3. This is employed in our calculations of the energy bands and eigenfunctions. Section 4 is devoted to the calculation of the polarizability and dielectric function of BP showing its temperature dependence and their anisotropic properties as a consequence of its band structure. Impurity shielding by BP is discussed in Section 5 and we summarize our important results in Section 6.

2. Surface response function for a pair of 2D layers

Let us consider a heterostructure whose surface is in the xy-plane and suppose that $r_\parallel$ denotes the corresponding in-plane translation vector. At time t, an external potential $\tilde{\phi}_{ext}(q, \omega)$ with wave vector q and frequency ω will give rise to an induced potential which, outside the structure, can be written as

$$\phi_{ind}(\mathbf{r}_\parallel, t) = -\int \frac{d^2\mathbf{q}}{(2\pi)^2} \int_{-\infty}^{\infty} d\omega \tilde{\phi}_{ext}(q, \omega) e^{i(\mathbf{q}\cdot\mathbf{r}_\parallel - \omega t)} g(\mathbf{q}, \omega) e^{-qz}. \tag{1}$$

This equation defines the surface response function $g(\mathbf{q}, \omega)$. It has been implicitly assumed that the external potential ϕ_{ext} is so weak that the medium responds linearly to it.

The quantity $\mathrm{Im}[g(\mathbf{q}, \omega)]$ can be identified with the power absorption in the structure due to electron excitation induced by the external potential. The total potential in the vicinity of the surface ($z \approx 0$), is given by

$$\phi(\mathbf{r}_\parallel, t) = \int \frac{d^2\mathbf{q}}{(2\pi)^2} \int_{-\infty}^{\infty} d\omega (e^{qz} - g(\mathbf{q}, \omega) e^{-qz}) e^{i(\mathbf{q}\cdot\mathbf{r}_\parallel - \omega t)} \tilde{\phi}_{ext}(\mathbf{q}, \omega) \tag{2}$$

which takes account of nonlocal screening of the external potential.

2.1 Model for phosphorene layer

In this section, we present the surface response function we calculated for a structure which consists of a pair of 2D layers in contact with a dielectric medium,

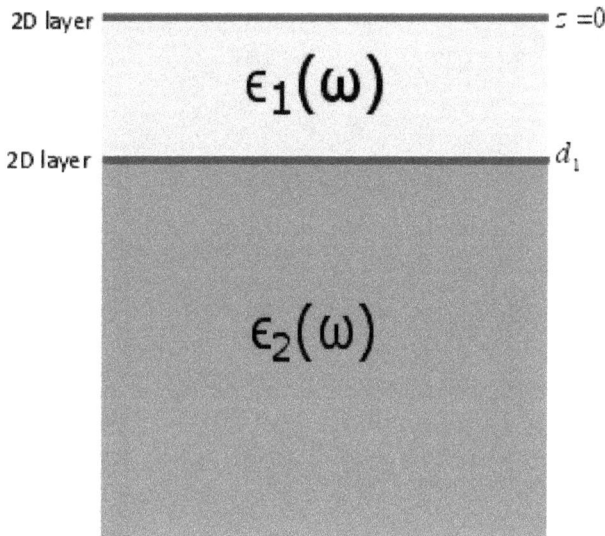

Figure 1.
(Color online) Schematic illustration of a hybrid structure consisting of a pair of 2D layers separated by distance d_1. The background materials are labeled by dielectric functions $\epsilon_1(\omega)$ and $\epsilon_2(\omega)$.

as shown in **Figure 1**. One of the 2D layers is at the top and the other is encapsulated by materials with dielectric constants $\epsilon_1(\omega)$, with thickness d_1, and $\epsilon_2(\omega)$, of semi-infinite thickness. Calculation shows that the surface response function is given by [29, 30].

$$g(\mathbf{q}, \omega) = \frac{\mathcal{N}(\mathbf{q}, \omega)}{\mathcal{D}(\mathbf{q}, \omega)}, \tag{3}$$

where

$$\mathcal{N}(\mathbf{q}, \omega) \equiv e^{2d_1q}\{q\epsilon_0(\epsilon_1(\omega) - 1) - \chi_1(\mathbf{q}, \omega)\}\{q\epsilon_0(\epsilon_1(\omega) + \epsilon_2(\omega)) - \chi_2(\mathbf{q}, \omega)\} \\ -\{q\epsilon_0(\epsilon_1(\omega) + 1) + \chi_1(\mathbf{q}, \omega)\}\{q\epsilon_0(\epsilon_1(\omega) - \epsilon_2(\omega)) + \chi_2(\mathbf{q}, \omega)\}, \tag{4}$$

and

$$\mathcal{D}(\mathbf{q}, \omega) \equiv e^{2d_1q}\{q\epsilon_0(\epsilon_1(\omega) + 1) - \mathcal{X}_1(\mathbf{q}, \omega)\}\{q\epsilon_0(\epsilon_1(\omega) + \epsilon_2(\omega)) - \mathcal{X}_2(\mathbf{q}, \omega)\} \\ -\{q\epsilon_0(\epsilon_1(\omega) - 1) + \mathcal{X}_1(\mathbf{q}, \omega)\}\{q\epsilon_0(\epsilon_1(\omega) - \epsilon_2(\omega)) + \mathcal{X}_2(\mathbf{q}, \omega)\}. \tag{5}$$

In this notation, $\mathbf{q}$ is the in-plane wave vector, ω is the frequency and $\chi_1(\mathbf{q}, \omega)$ and $\chi_2(\mathbf{q}, \omega)$ are the 2D layer susceptibilities.

When we take the limit $d_1 \rightarrow 0$, i.e., the separation between the two layer is small, the ϵ_1 drops out and we have the following result for the surface response function corresponding to the structure in **Figure 2**

$$g(\mathbf{q}, \omega) = 1 - \frac{1}{\frac{1 + \epsilon_2(\omega)}{2} - \frac{\chi_1(\mathbf{q}, \omega) + \chi_2(\mathbf{q}, \omega)}{2q\epsilon_0}}. \tag{6}$$

Here, the dispersion equation which is given by the zeros of the denominator $\epsilon(\mathbf{q}, \omega)$ of the second term is expressed in terms of the 'average' susceptibility for the two layers. Clearly, this dispersion equation is that for a 2D layer of the Stern form where we make the identification $\chi \rightarrow e^2 \Pi^{(0)}$ in terms of the polarizability. This

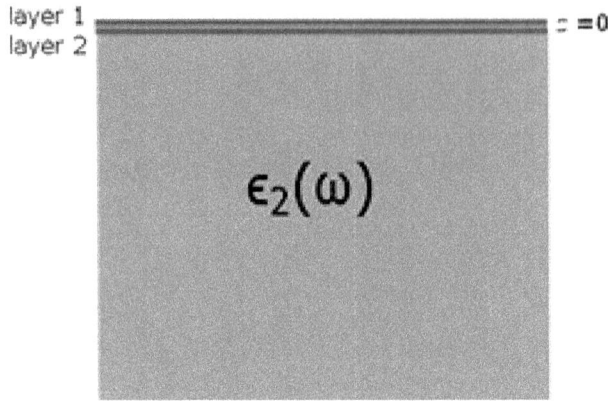

Figure 2.
(Color online) Schematic representation of a structure consisting of a pair of 2D layers which are infinitesimally close. There is vacuum above the layers and a dielectric below.

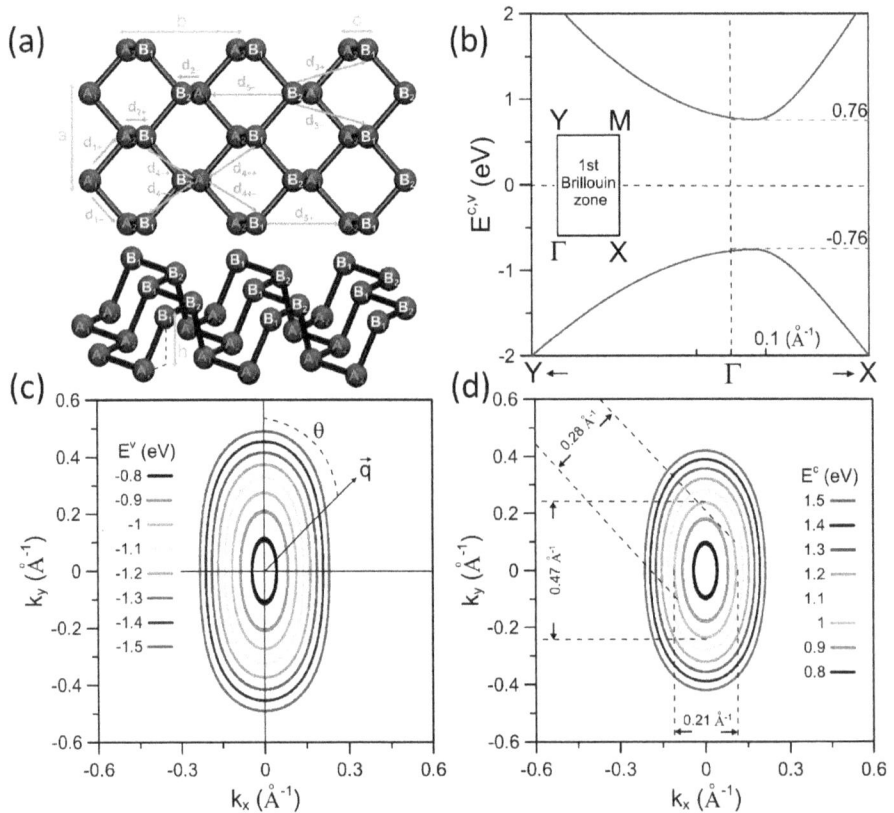

Figure 3.
(Color online) The (a) top view and side view of crystal structure for BP and (b) its band structure. The constant-energy diagrams are presented for (c) valence and (d) conduction bands. The values of 2 k_F for different θ are given in (d).

result in Eq. (6) clearly illustrates that for the buckled BP structure shown in **Figure 3**, the dielectric function can be treated as that for a single layer whose susceptibility arises from a *combination* of two rows of atoms making up the layer.

Our calculation can easily be generalized to the case when the monolayer is embedded above and below by the same thick dielectric material (dielectric constant ϵ_b) which corresponds to the free-standing situation which we consider below. For this, we have $\epsilon(\mathbf{q}, \omega) = \epsilon_b - e^2/(2\epsilon_0 q)\Pi^{(0)}(\mathbf{q}, \omega)$, expressed in terms of the 2D layer polarizability $\Pi^{(0)}(\mathbf{q}, \omega)$.

3. Model Hamiltonian

Phosphorene is treated as a single layer of phosphorus atoms arranged in a puckered orthorhombic lattice, as shown in **Figure 3(a)**. It contains two atomic layers of A and B atoms and two kinds of bonds for in-plane and inter-plane P–P connections with different bond lengths. The low-lying electronic structure can be described by a tight-binding Hamiltonian, which is a 4×4 matrix within the basis (A_1, A_2, B_1, B_2), of the form

$$\begin{bmatrix} 0 & T_1 + T_3^* & T_4 & T_{2-} + T_{5-}^* \\ T_1^* + T_3 & 0 & T_{2+} + T_{5+}^* & T_4 \\ T_4 & T_{2+}^* + T_{5+} & 0 & T_1 + T_3^* \\ T_{2-}^* + T_{5-} & T_4 & T_1^* + T_3 & 0 \end{bmatrix}.$$

Here, we consider up to five nearest atomic interactions through five independent terms of T_i with $i = 1, 2, 3, 4, 5$. These terms are given by the following expressions.

$$\begin{cases} T_1 = t_1 e^{i\mathbf{k}\cdot(\mathbf{d}_{1+} + \mathbf{d}_{1-})} \\ T_{2\pm} = t_2 e^{i\mathbf{k}\cdot\mathbf{d}_{2\pm}} \\ T_3 = t_3 e^{i\mathbf{k}\cdot(\mathbf{d}_{3+} + \mathbf{d}_{3-})} \\ T_4 = t_4 e^{i\mathbf{k}\cdot\left(\mathbf{d}_{4++} + \mathbf{d}_{4+-} + \mathbf{d}_{4-+} + \mathbf{d}_{4--}\right)} \\ T_{5\pm} = t_5 e^{i\mathbf{k}\cdot\mathbf{d}_{5\pm}}. \end{cases} \tag{7}$$

In this notation, t_m ($m = 1, 2, 3, 4, 5$) are the hopping integrals, corresponding to the atomic interactions. They have been optimized as ($t_1 = -1.220$, $t_2 = 3.665$, $t_3 = -0.205$, $t_4 = -0.105$, $t_5 = -0.055$) in order to reproduce the energy bands obtained by the density functional theory (DFT) calculations [31–33]. Also, $\vec{d}_{m\pm}$ are the vectors connecting the lattice sites which can be written as

$$\begin{cases} d_{1\pm} = (b/2 - c, \pm a/2, 0) \\ d_{2\pm} = (\pm c, 0, h) \\ d_{3\pm} = (b/2 + c, \pm a/2, 0) \\ d_{4\pm} = (\pm b/2, \pm a/2, h) \\ d_{5\pm} = \{\pm (b - c), 0, -h\}, \end{cases} \tag{8}$$

where $a = 3.314$Å, $b = 4.376$Å, $c = 0.705$Å, and $h = 2.131$Å are the distances between the BP atoms [34, 35], as illustrated in **Figure 3(a)**.

The valence and conduction energy bands present strong anisotropic behaviors, as illustrated by the energy bands in **Figure 3(b)** and the constant-energy loops in **Figure 3(c)** and **(d)**. As a result, the polarizability and dielectric function are shown to be strongly dependent on the direction of the transferred momentum $\mathbf{q}$.

4. Dielectric function

When monolayer BP is perturbed by an external time-dependent Coulomb potential, all the valence and conduction electrons will screen this field and therefore create the charge redistribution. The effective potential between two charges is the sum of the external potential and the induced potential due to screening charges. The dynamical dielectric function, within the random-phase approximation (RPA), is given by [36].

$$\epsilon(\mathbf{q}, \omega) = \epsilon_b - V_q \sum_{s,\, s'=\alpha,\, \beta} \sum_{h,\, h'=c,\, v} \int_{1stBZ} \frac{dk_x dk_y}{(2\pi)^2} |\langle s'; h'; \mathbf{k} + \mathbf{q} | e^{i\mathbf{q} \cdot \mathbf{r}} | s; h; \mathbf{k} \rangle|^2$$

$$\times \frac{f\left(E^{s',h'}(\mathbf{k} + \mathbf{q})\right) - f\left(E^{s,h}(\mathbf{k})\right)}{E^{s',h'}(\mathbf{k} + \mathbf{q}) - E^{s,h}(\mathbf{k}) - (\omega + i\Gamma)}. \tag{9}$$

Here, the π-electronic excitations are described in terms of the transferred momentum $\mathbf{q}$ and the excitation frequency ω. $\epsilon_b = 2.4$ the background dielectric constant, $V_q = 2\pi e^2/(\epsilon_s q)$ the 2D Fourier transform of the bare Coulomb potential energy ($\epsilon_s = 4\pi\epsilon_0$), and Γ the energy width due to various de-excitation mechanisms. $f(E) = 1/\{1 + \exp{[(E - \mu)k_B T]}\}$ the Fermi-Dirac distribution in which k_B is the Boltzmann constant and μ the chemical potential corresponding to the highest occupied state energy (middle energy of band gap) in the (semiconducting) metallic systems at T = 0 K.

Figure 4(a) and **(b)** shows the directional/θ-dependence of the static polarization function $\Pi^{(0)}(0, \mathbf{q})$, in which θ defines the angle between the direction of $\mathbf{q}$ and the unit vector $\hat{k}_y$. For arbitrary θ, the polarization function at lower ($q \leq 0.2\ (1/\text{Å})$)

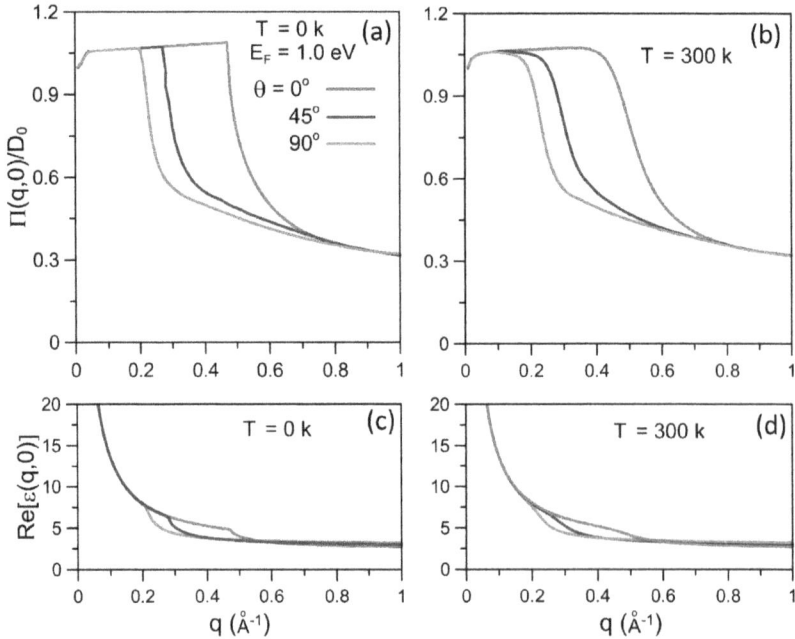

Figure 4.
(Color online) The static polarizability for BP as a function of wave vector for different directions of the transferred momentum **q** at (a) absolute zero and (b) room temperatures. Plots (c) and (d) correspond to the static dielectric function of BP at T = 0 and 300 K, respectively.

and higher $(q \geq 0.7 \, (1/\text{Å}))$ transferred momentum remains unchanged. In general, $\Pi^{(0)}(0, \mathbf{q})$ falls off rapidly beyond a critical value of q ($2k_F$) which depends on θ. For increasing θ from 0 to 90°, the specific values are getting larger, as shown in **Figure 4(a)**. This means that the polarizability is stronger for $0.2 \leq q \leq 0.7 \, (1/\text{Å})$. The main features of the polarizability for BP are quite similar to those for the 2D electron gas, but different with those for graphene. Temperature has an effect on the polarization function which is demonstrated in **Figure 4(b)**. At room temperature, $\Pi^{(0)}(0, \mathbf{q})$ exhibits a shoulder-like structure near the critical values of q instead of step-like structure at T = 0.

Plots of the static dielectric function of BP for various values of θ are presented in **Figure 4(c)** and **(d)** at absolute zero and room temperatures, respectively. In the range of $0.2 \leq q \leq 0.5 \, (1/\text{Å})$, there is a clear dependence of the dielectric function on the direction of the transferred momentum q. The Re $\epsilon(0, q)$ is higher with the growth of θ. The introduction of finite temperature smoothens the q-dependent Re $\epsilon(0, q)$, as shown in **Figure 4(d)** for T = 30 K.

5. Impurity shielding

Starting with Eq. (2), we obtain the static screening of the potential on the surface at $z = 0$ due to an impurity with charge $Z_0^* e$ located at distance z_0 above the surface of BP as

$$
\phi(\mathbf{r}_{\parallel}, \omega = 0) = \frac{Z_0^* e}{2\pi\epsilon_0} \int_0^{\infty} dq \int_0^{2\pi} d\theta e^{iqr\cos\theta}[1 - g(\mathbf{q}, \omega = 0)]e^{-qz_0}
$$
$$
= \frac{Z_0^* e}{2\pi\epsilon_0} \int_0^{\infty} dq \int_0^{2\pi} d\theta \frac{e^{iqr\cos\theta - qz_0}}{\epsilon(\mathbf{q}, \omega = 0)}.
$$
(10)

By employing the generalized form of Eq. (6) for free-standing BP in Eq. (10), we have computed the screened impurity potential. The screened potentials for various z_0's are shown in **Figure 5** at absolute zero temperature and Fermi energy

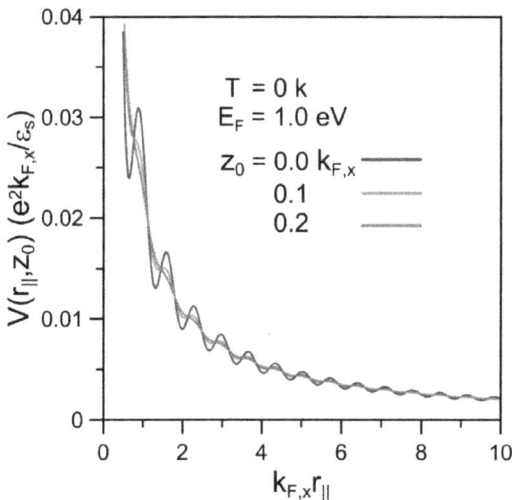

Figure 5.
(Color online) The screened impurity potential in units of $e^2 k_{F,x}/(\epsilon_s)$ is plotted as a function of $k_{F,x} r_{\parallel}$ for the chosen parameters in the figure.

E_F = 1.0 eV. There exist Friedel oscillations for sufficiently small z_0. Such oscillations might be smeared out for larger z_0, e.g., the green and red curves. It is noticed that for E_F = 1.0 eV, the room temperature of 300 K which is much smaller than the Fermi temperature (10,000 K) does not have significant effect on the screened potential. Apparently, $V(r_\parallel, z_0)$ at T = 0 and 300 K (not shown) are almost equivalent.

6. Concluding remarks and summary

The energy band structure of BP, calculated using the tight-binding method, is anisotropic and so are its polarizability, dielectric function and screened potential. To illustrate these facts, we have presented numerical results for the polarizability in the x and y directions for a range of doping concentrations. The $\mathrm{Re}[\epsilon(\mathbf{q}, \omega = 0)]$ of the static dielectric function for BP also reveals some interesting characteristics. At absolute zero temperature (T = 0) and with free carrier density corresponding to chosen Fermi energy E_F, we have presented numerical results for $\mathrm{Re}[\epsilon(\mathbf{q}, \omega = 0)]$ in different directions of the transferred momentum $\mathbf{q}$. When q is larger than a critical value which is twice the Fermi momentum k_F, our calculations show that Re $[\epsilon(\mathbf{q}, \omega = 0)]$ becomes substantially dependent on the direction of $\mathbf{q}$. We also discuss the case at room temperature (T = 300 K). These results are in agreement with those reported by other authors. We employ our data to determine the static shielding of an impurity in the vicinity of phosphorene.

Acknowledgements

G.G. would like to acknowledge the support from the Air Force Research Laboratory (AFRL) through Grant #12530960.

Conflict of interest

All the authors declare that they have no conflict of interest.

Author details

Po Hsin Shih[1], Thi Nga Do[2], Godfrey Gumbs[3,4*] and Dipendra Dahal[3]

1 Department of Physics, National Cheng Kung University, Tainan, Taiwan

2 Institute of Physics, Academia Sinica, Taipei, Taiwan

3 Department of Physics and Astronomy, Hunter College of the City University of New York, New York, NY, USA

4 Donostia International Physics Center (DIPC), San Sebastian, Basque Country, Spain

*Address all correspondence to: ggumbs@hunter.cuny.edu

IntechOpen

References

[1] Novoselov KS, Geim AK, Morozov SV, Jiang D, Zhang Y, Dubonos SV, et al. Electric field effect in atomically thin carbon films. Science. 2004;**306**: 666-669. DOI: 10.1126/science.1102896

[2] Dbbelin M, Ciesielski A, Haar S, Osella S, Bruna M, Minoia A, et al. Light-enhanced liquid-phase exfoliation and current photoswitching in graphene-azobenzene composites. Nature Communications. 2016;**7**:11090. DOI: 10.1038/ncomms11090

[3] Kim KS, Zhao Y, Jang H, Lee SY, Kim JM, Kim KS, et al. Large-scale pattern growth of graphene films for stretchable transparent electrodes. Nature. 2009;**457**:706-710. DOI: 10.1038/nature07719

[4] Tao L, Cinquanta E, Chiappe D, Grazianetti C, Fanciulli M, Dubey M, et al. Silicene field-effect transistors operating at room temperature. Nature Nanotechnology. 2015;**10**:227-231. DOI: 10.1038/nnano.2014.325

[5] Vogt P, De Padova P, Quaresima C, Avila J, Frantzeskakis E, Asensio MC, et al. Silicene: Compelling experimental evidence for Graphenelike two-dimensional silicon. Physical Review Letters. 2012;**108**:155501. DOI: 10.1103/PhysRevLett.108.155501

[6] Li L, Yu Y, Ye GJ, Ge Q, Ou X, Wu H, et al. Black phosphorus field-effect transistors. Nature Nanotechnology. 2014;**9**:372-377. DOI: 10.1038/nnano.2014.35

[7] Yasaei P, Kumar B, Foroozan T, Wang C, Asadi M, Tuschel D, et al. High-quality black phosphorus atomic layers by liquid-phase exfoliation. Advanced Materials. 2015;**27**:1887-1892. DOI: 10.1002/adma.201405150

[8] Li L, Lu S, Pan J, Qin Z, Wang Y, Wang Y, et al. Buckled Germanene formation on Pt(111). Advanced Materials. 2014;**26**:4820-4824. DOI: 10.1002/adma.201400909

[9] Derivaz M, Dentel D, Stephan R, Hanf M-C, Mehdaoui A, Sonnet P, et al. Continuous Germanene layer on Al (111). Nano Letters. 2015;**15**:2510-2516. DOI: 10.1021/acs.nanolett.5b00085

[10] Ji J, Song X, Liu J, Yan Z, Huo C, Zhang S, et al. Two-dimensional antimonene single crystals grown by van der Waals epitaxy. Nature Communications. 2016;**7**:13352. DOI: 10.1038/ncomms13352

[11] Ares P, Aguilar Galindo F, Rodríguez San Miguel D, Aldave DA, Díaz Tendero S, Alcamí M, et al. Antimonene: Mechanical isolation of highly stable antimonene under ambient conditions. Advanced Materials. 2016; **28**:6515-6515. DOI: 10.1002/adma.201670209

[12] Zhu F, Chen W, Xu Y, Gao C, Guan D, Liu C, et al. Epitaxial growth of two-dimensional stanene. Nature Materials. 2015;**14**:1020-1025. DOI: 10.1038/nmat4384

[13] Hirahara T, Nagao T, Matsuda I, Bihlmayer G, Chulkov EV, Koroteev YM, et al. Role of spin-orbit coupling and hybridization effects in the electronic structure of ultrathin Bi films. Physical Review Letters. 2006;**97**: 146803. DOI: 10.1103/PhysRevLett.97.146803

[14] Hirahara T, Shirai T, Hajiri T, Matsunami M, Tanaka K, Kimura S, et al. Role of quantum and surface-state effects in the bulk Fermi-level position of ultrathin Bi films. Physical Review Letters. 2015;**115**:106803. DOI: 10.1103/PhysRevLett.115.106803

[15] Hirahara T, Fukui N, Shirasawa T, Yamada M, Aitani M, Miyazaki H, et al.

Atomic and electronic structure of ultrathin Bi(111) films grown on $Bi_2Te_3(111)$ substrates: Evidence for a strain-induced topological phase transition. Physical Review Letters. 2012;**109**:227401. DOI: 10.1103/PhysRevLett.109.227401

[16] Yang F, Miao L, Wang ZF, Yao M-Y, Zhu F, Song YR, et al. Spatial and energy distribution of topological edge states in single Bi(111) bilayer. Physical Review Letters. 2012;**109**:016801. DOI: 10.1103/PhysRevLett.109.016801

[17] Wang ZF, Yao M-Y, Ming W, Miao L, Zhu F, Liu C, et al. Creation of helical Dirac fermions by interfacing two gapped systems of ordinary fermions. Nature Communications. 2013;**4**:1384. DOI: 10.1038/ncomms2387

[18] Sabater C, Gosálbez-Martínez D, Fernández-Rossier J, Rodrigo JG, Untiedt C, Palacios JJ. Topologically protected quantum transport in locally exfoliated bismuth at room temperature. Physical Review Letters. 2013;**110**:176802. DOI: 10.1103/PhysRevLett.110.176802

[19] Malcolm JD, Nicol EJ. Frequency-dependent polarizability, plasmons, and screening in the two-dimensional pseudospin-1 dice lattice. Physical Review B. 2016;**93**:165433. DOI: 10.1103/PhysRevB.93.165433

[20] Liu H, Neal AT, Zhu Z, Luo Z, Xu X, Tomnek D, et al. Phosphorene: An unexplored 2D semi-conductor with a high hole mobility. ACS Nano. 2014;**8**: 4033-4041. DOI: 10.1021/nn501226z

[21] Brent JR, Savjani N, Lewis EA, Haigh SJ, Lewis DJ, O'Brien P. Production of few-layer phosphorene by liquid exfoliation of black phosphorus. Chemical Communications. 2014;**50**: 13338-13341. DOI: 10.1039/C4CC05752J

[22] Kang J, Wood JD, Wells SA, Lee J-H, Liu X, Chen K-S, et al. Solvent exfoliation of electronic-grade, two-dimensional black phosphorus. ACS Nano. 2015;**9**: 3596-3604. DOI: 10.1021/acsnano.5b01143

[23] Lange S, Schmidt P, Nilges T. Au3SnP7@black phosphorus: An easy access to black phosphorus. Inorganic Chemistry. 2007;**46**:4028-4035. DOI: 10.1021/ic062192q

[24] Nilges T, Kersting M, Pfeifer T. A fast low-pressure transport route to large black phosphorus single crystals. Journal of Solid State Chemistry. 2008; **181**:1707-1711. DOI: 10.1016/j.jssc.2008.03.008

[25] Köpf M, Eckstein N, Pfister D, Grotz C, Krüger I, Greiwe M, et al. Access and in situ growth of phosphorene-precursor black phosphorus. Journal of Crystal Growth. 2014;**405**:6-10. DOI: 10.1016/j.jcrysgro.2014.07.029

[26] Rudenko AN, Katsnelson MI. Quasiparticle band structure and tight-binding model for single- and bilayer black phosphorus. Physical Review B. 2014;**89**:201408. DOI: 10.1103/PhysRevB.89.201408

[27] Berman OL, Gumbs G, Kezerashvili RY. Bose-Einstein condensation and superfluidity of dipolar excitons in a phosphorene double layer. Physical Review B. 2017;**96**:014505. DOI: 10.1103/PhysRevB.96.014505

[28] Low T, Rodin AS, Carvalho A, Jiang Y, Wang H, Xia F, et al. Tunable optical properties of multilayer black phosphorus thin films. Physical Review B. 2014;**90**:075434. DOI: 10.1103/PhysRevB.90.075434

[29] Gumbs G, Huang D. Properties of Interacting Low-Dimensional Systems. 1st ed. New York: Wiley; 2011. 113 p. DOI: 10.1002/9783527638154

[30] Dahal D, Gumbs G, Huang D. Effect of strain on plasmons, screening, and

energy loss in graphene/substrate contacts. Physical Review B. 2018;**98**: 045427. DOI: 10.1103/ PhysRevB.98.045427

[31] Heyd J, Scuseria GE, Ernzerhof M. Hybrid functionals based on a screened Coulomb potential. The Journal of Chemical Physics. 2003;**118**:8207-8215. DOI: 10.1063/1.1564060

[32] Heyd J, Scuseria GE, Ernzerhof M. Erratum. Hybrid functionals based on a screened Coulomb potential. The Journal of Chemical Physics. 2006;**124**: 219906. DOI: 10.1063/1.2204597

[33] Pez CJ, DeLello K, Le D, Pereira ALC, Mucciolo ER. Disorder effect on the anisotropic resistivity of phosphorene determined by a tight-binding model. Physical Review B. 2016; **94**:165419. DOI: 10.1103/ PhysRevB.94.165419

[34] Takao Y, Asahina H, Morita A. Electronic structure of black phosphorus in tight binding approach. Journal of the Physical Society of Japan. 1981;**50**:3362-3369. DOI: 10.1143/ JPSJ.50.3362

[35] Osada T. Edge state and intrinsic hole doping in bilayer phosphorene. Journal of the Physical Society of Japan. 2014;**84**:013703. DOI: 10.7566/ JPSJ.84.013703

[36] Shung KW-K. Dielectric function and plasmon structure of stage-1 intercalated graphite. Physical Review B. 1986;**34**:979-993. DOI: 10.1103/ PhysRevB.34.979

Chapter 6

MoS₂ Thin Films for Photo-Voltaic Applications

Manuel Ramos, John Nogan, Manuela Ortíz-Díaz,
José L Enriquez-Carrejo, Claudia A Rodriguez-González,
José Mireles-Jr-Garcia, Roberto Carlos Ambrosio-Lazáro,
Carlos Ornelas, Abel Hurtado-Macias, Torben Boll,
Delphine Chassaing and Martin Heilmaier

Abstract

The low dimensional chalcogenide materials with high band gap of ~1.8 eV, specially molybdenum di-sulfide (MoS_2), have been brought much attention in the material science community for their usage as semiconducting materials to fabricate low scaled electronic devices with high throughput and reliability, this includes also photovoltaic applications. In this chapter, experimental data for MoS_2 material towards developing the next generation of high-efficiency solar cells is presented, which includes fabrication of ~100 nm homogeneous thin film over silicon di-oxide (SiO_2) by using radio frequency sputtering at 275 W at high vacuum~10^{-9} from commercial MoS_2 99.9% purity target. The films were studied by means of scanning and transmission electron microscopy with energy disperse spectroscopy, grazing incident low angle x-ray scattering, Raman spectroscopy, atomic force microscopy, atom probe tomography, electrical transport using four-point probe resistivity measurement as well mechanical properties utilizing nano-indentation with continuous stiffness mode (CSM) approach. The experimental results indicate a vertical growth direction at (101)-MoS_2 crystallites with stacking values of 7-laminates along the (002)-basal plane; principal Raman vibrations at E^1_{2g} at 378 cm^{-1} and A^1_g at 407 cm^{-1}. The hardness and elastic modulus values of H = 10.5 ± 0.1 GPa and E = 136 ± 2 GPa were estimated by CSM method from 0 to 90 nm of indenter penetration; as well transport measurements from −3.5 V to +3.5 V indicating linear Ohmic behavior.

Keywords: thin film, electron microscopy, MoS_2 sputtering, harness, elastic modulus, x-ray diffraction, electrical transport, focus ion-beam, atom probe tomography

1. Introduction

Layered chalcogenide materials have been of high relevance since almost 40 years for their diverse applications such as tribology [1], chemical catalysis [2] and nowadays as semiconductors towards development of high-throughput and energy efficient transistors and devices [3, 4]. MoS_2 is a two-dimensional material with a band gap ranging between 0.9 and 1.8 eV as calculated theoretically by first principles methods and as measured experimentally by Kam & Parkinson using photo-spectroscopy as a function of crystal orientation [5, 6]. The crystal structure of MoS$_2$ is hexagonal with space group R3m ($a = b = 3.16$ Å and $c = 18.41$ Å), having d-bonded layers of S-Mo-S along a-b plane which are stacked along c-axis by weak Van der Waals forces with 6.2 Å of separation within layers [7]. The crystal structure was studied using electron microscopy techniques as described by Chianelli et al. who were able to observe its layered structure [8]. However, electron beam dosage during electron microscopy studies plays an important role to avoid any structural damage as described by Ponce et al. when using TEM technique who concluded high-resolution imaging at operational voltages near ~80 kV [9] to be possible. By *in-situ* TEM, Helveg et al. were able to synthesize small clusters of MoS$_2$ from molybdenum oxide and hydrogen sulfide gases at beam radiation dosage of 100 e^-/Å^{2}s [10]. The mechanical properties were studied by Casillas et al. achieving an atomistic observation of a resilient nature on MoS$_2$ laminates at 8GPa of external applied pressure and its mechanical recovery during *in-situ* AFM on TEM sample holder [11]. Applying atomic force microscopy (AFM), Bertolazzi et al. determined a Young modulus values of 270GPa ± 100GPa and fracture strength of 16~30GPa in MoS$_2$ layers as suspended in patterned silicon wafers [12, 14], and Castellanos-Gomez et al. estimated an average Young modulus $E = 330$GPa in suspended MoS$_2$ sheets over patterned silicon wafer [13]. The mechanical properties were studied by density functional theory and molecular dynamics, Jiang et al. calculated a theoretical Poisson's ratio value of $v = 0.29$ applying Stillinger-Weber potential [15]. The reactive empirical bond-order (REBO) potential was used by Li et al. to understand structural effects at chemical bonding within S-Mo-S layers, their findings indicate induced vacancies on the basal plane can influence Poisson's ratio values [16]. The atom probe tomography enables the chemical understanding with three-dimensional spatial resolution and was applied to determine dopants, contamination and ionic distribution within semiconducting matrix [17], Singh et al. used APT technique to determine distribution of Ti over MoS$_2$ matrix [18]. Regarding electrical transport, Lia et al. [4] and Samuel et al. [38] performed transport electrical measurements encountering a linear ohmic behavior in MoS$_2$. This chapter covers mechanical, electrical and microstructure characterization by electron microscopy, low angle x-ray, atom probe tomography and CSM-nanoindentation to obtain information about crystal growth, elastic modulus (E), hardness (H) and electrical transport on MoS$_2$ films.

2. Experimental methods and results

2.1 RF sputtering

The Molybdenum di-Sulfide (MoS$_2$) films were fabricated with a high vacuum Kurt J. Lesker© PVD 75 machine; applying RF-sputtering at a rate of 2.26 Å/sec at 275 W of plasma power over 4″-diameter silicon oxide (SiO$_2$) wafers. The films were deposit from commercial MoS$_2$ 99.9% targets (Kurt J. Lesker). By using dwell time of 300 seconds a film thickness value of ~100 nm was achieved as indicated by profilometry measurements, **Figure 1E**.

2.2 Scanning electron microscopy

The film morphology and crystallographic structure were investigated using scanning and high-resolution transmission electron microscopy (SEM, TEM). SEM

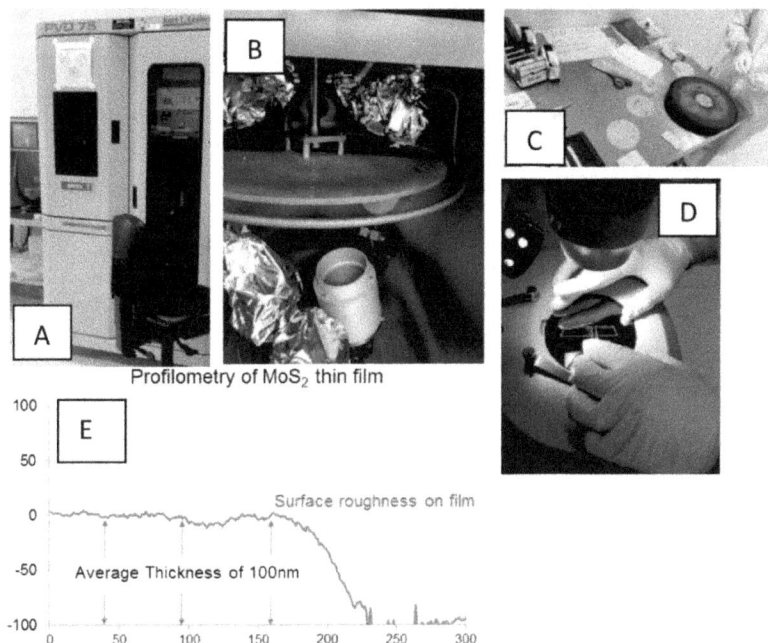

Figure 1.
Collage of photographic images taken at the Center for Integration of nanotechnologies-Albuquerque, NM. (A) High-vacuum sputtering device, overview. (B) RF-magnetrons with MoS₂ target mounted. (C) Table showing the sample mounting. (D) Optical inspection of the films. (E) Profilometer measurements indicating the film thickness ~ 100 nm for 300 seconds a rate of 2.26 Å/sec.

was performed in a Hitachi® SU5500 unit, equipped with Energy-dispersive X-ray spectroscopy (EDS) unit and operated at 30 kV with 8A of current to avoid surface damage on the film. Observations indicate a high-degree of porosity and vertically aligned MoS_2 film matrix, as presented in **Figure 2A–C** which is in agreement with Kong et al. [18]. EDS analysis reveals the two main signals that correspond to Sulfur-K_α and Molybdenum-L_α at 2.4 keV, as presented in **Figure 2E**, in agreement with Lince & Fleischauer [20].

2.3 Transmission electron microscopy and atom probe tomography

The microstructure of MoS_2 thin matrix was also studied using *Scanning Electron Transmission Microscopy* (STEM) using a Cs-corrected 2200-JEOL, with STEM unit, equipped with a high-angle annular dark-field (HAADF) detector, X-Twin lenses and CCD camera. A lamella was prepared using Focus-Ion Beam model JEOL JEM 9320 at 30 kV and 25 mA, MoS_2 film surface coated with gold and gallium. *Atom Probe Tomography* (APT) was performed on Cameca® LEAP 4000X high-resolution system in laser pulse mode (wavelength ~355 nm), measurements were taken at 60 K with evaporation rate of 0.5 and laser frequency of 100 kHz, laser beam was set to 70pJ/V, all data was reconstructed using IVAS© 3.6.10a package. The samples were prepared using focus ion beam FEI Strata dual-beam instrument coupled with micromanipulator Oxford® Omniprobe® 200 by lift-out method as described by Szász et al. [21]. MoS_2 film surface was protected using platinum layer, and cuts were done at 30 kV at 260pA gun power. Using both techniques, it was possible to determine chemical composition, and spatial resolution of S-Mo-S distribution along film matrix, stacking and orientation, as presented in **Figure 3**. In the image the top part corresponds to MoS_2 and uppermost bright layer is due to gold gallium coating.

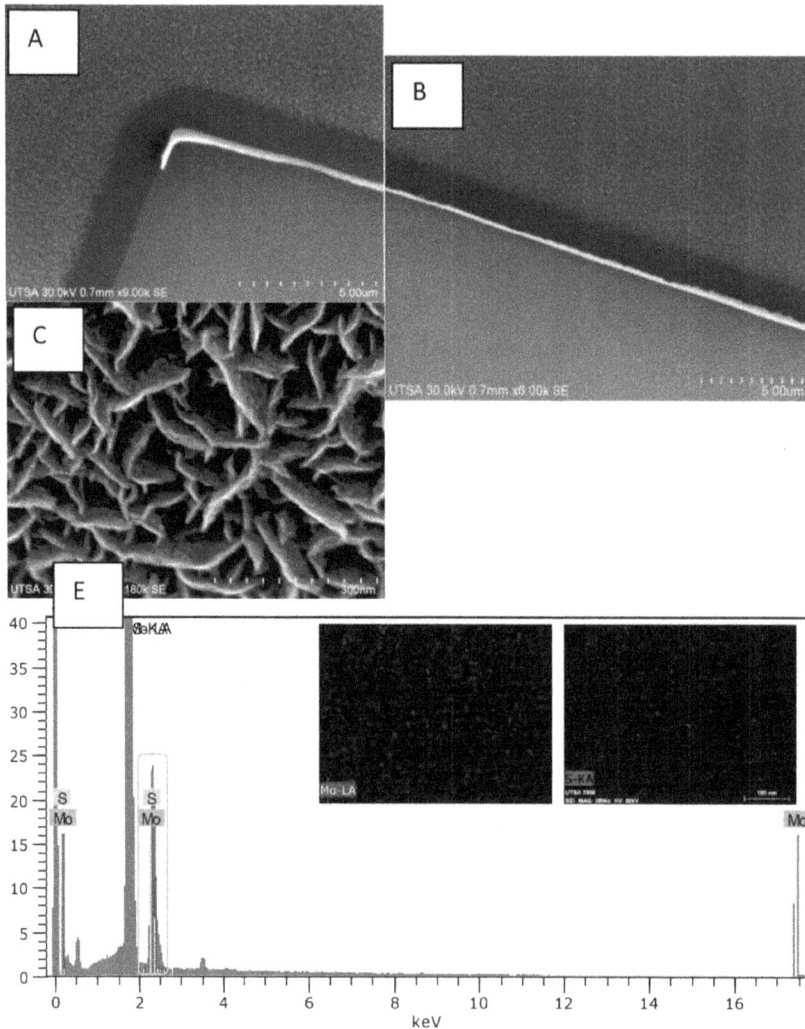

Figure 2.
(**A, B**) *Scanning electron micrograph of MoS₂ film matrix at magnification of 6000x.* (**C**) *SEM image able to observe laminates vertically aligned at magnification of 200,000x.* (**E**) *Energy disperse spectrum from surface to MoS₂ film matrix to determine chemical composition.*

The atom probe tomography (APT) is a technique used to understand in a three-dimensional reconstruction with high-spatial resolution the chemical distribution and composition as indicated by Kelly & Miller [17]. A sample is placed in the main APT chamber to undergo an ionizing evaporation process at a high electric field triggered by a laser pulse; the potential energy of an atom at the sample surface, as caused by the applied voltage on the sample neV, is converted into kinetic energy $\sim 1/2mv^2$ in the vicinity of the tip. This relationship, in order to understand the mass-to-charge-state ratio m/n of evaporated ions, is given by Eq. (1); with n as number of electrons removed from the ion, e electron charge (-1.62×10^{-19}C), V total applied voltage, m is atomic mass and speed of atoms are given by conventional $v = d/t$, which is with good approximation constant, distance d and lastly t is the time of flight, as described by the schematic drawing taken from Kelly and Larson [23]. Short laser pulses (<1 ns) are used for APT and can field evaporated for almost any material regardless of its electrical conductivity as described by Kellogg et al. [24].

$$\frac{m}{n} = \frac{2e}{d^2}(V_{dc})t^2 \tag{1}$$

Nowadays, usage of APT to survey spatial distribution of atomistic species in semiconducting devices like n-doped metal-oxide field effect transistors [25] and Singh et al. applied with high success to titanium-MoS₂ and strontium oxide-MoS₂ films [18]. In this case, APT measurements were performed to understand the spatial distribution of MoS₂ film matrix. **Figure 4** illustrates the preparation of APT samples using a FIB (**Figures 5–9**).

2.4 Raman spectroscopy

The Raman spectroscopy was obtained using Alpha 300RA system equipped with a 532 nm Nd-YAG laser and a 100X 0.9 NA objective. The laser power was varied to avoid surface damage; with no additional sample preparation during study. Modes of vibration at $E^1_{2g} = 378$ cm^{-1} and $A^1_g = 407$ cm^{-1} are indicators of sulfur vibrations caused by dangling bongs on S-Mo-S chemical structure as indicated schematically **Figure 10** (insets).

2.5 Grazing incidence X-ray diffraction (GIXD)

X-ray diffraction was collected using a Panalytical X-Pert system with source of Cu$_{K\alpha}$ $\lambda = 1.41$ Å radiation. The grazing incidence angle was fixed at 0.5° with $20° < \theta < 80°$ and step size of 0.02° with a graphite flat crystal monochromator, described by Liu et al. while characterizing same layers of MoS₂ [26] and presented in **Figure 11**.

2.6 Nanoscale mechanical properties

The nanoscale mechanical properties were evaluated to obtain Elastic modulus (E) and Hardness (H) of MoS₂ thin films; this was possible using an Agilent nanoindenter model G200 coupled with a DCM II head instrument and Berkovich diamond indenter tip radius of 20 ± 5 nm, penetration depth limit of 400 nm, strain rate of 0.05 s^{-1}, and harmonic displacement and frequency of 1 nm and 75 Hz, Poisson's coefficient of $\nu = 0.22$. The equipment was calibrated using a standard

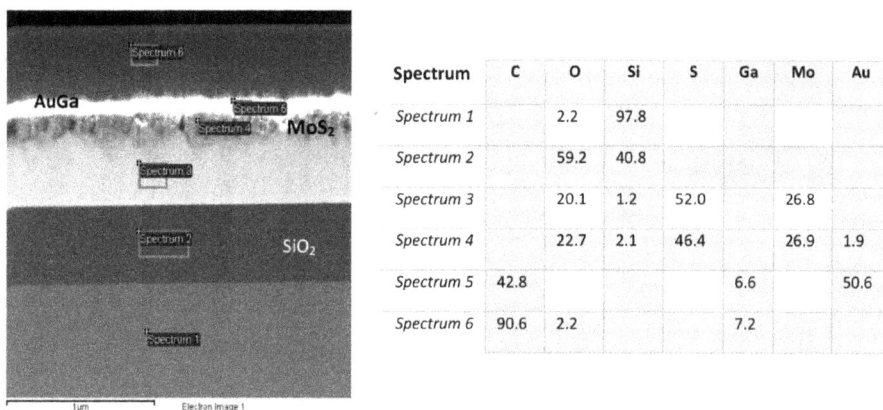

Spectrum	C	O	Si	S	Ga	Mo	Au
Spectrum 1		2.2	97.8				
Spectrum 2		59.2	40.8				
Spectrum 3		20.1	1.2	52.0		26.8	
Spectrum 4		22.7	2.1	46.4		26.9	1.9
Spectrum 5	42.8				6.6		50.6
Spectrum 6	90.6	2.2			7.2		

Figure 3.
Left: *Cross-sectional view of MoS₂ film on TEM.* **Right:** *Chemical composition at locations as indicated in the image (violet squares), obtained during TEM observations.*

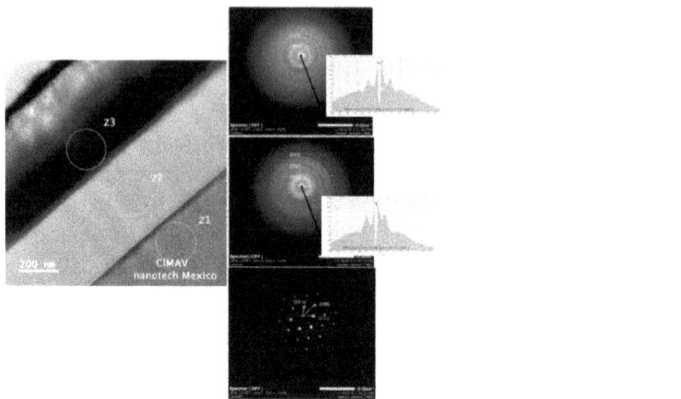

Zone	d-Theoretical (Å)	d-Experimental (Å)	Crystallographic Planes	Theoretical Angles	Experimental Angles
Z1	3.138	3.124	(111)	---	---
Z1	1.92	1.93	(220)	35.26	34.61
Z1	3.138	3.198	(11-1)	70.53	70.5
Z2	3.35	3.49	(101)	---	---
Z2	1.81	1.71	(112)		
Z2	1.20	1.2	(213)	---	---
Z3	3.35	3.452	(101)	---	---
Z3	1.81	1.718	(112)	---	---

Figure 4.
Left: *Scanning transmission electron micrograph of transversal section of MoS$_2$ film.* **Right:** *Selected area diffraction patterns for three different sites as indicated by red circles, the top part corresponds to textural MoS$_2$ matrix. Table indicates principal with (101) and (112) for MoS$_2$ in agreement with obtained by GDRX.*

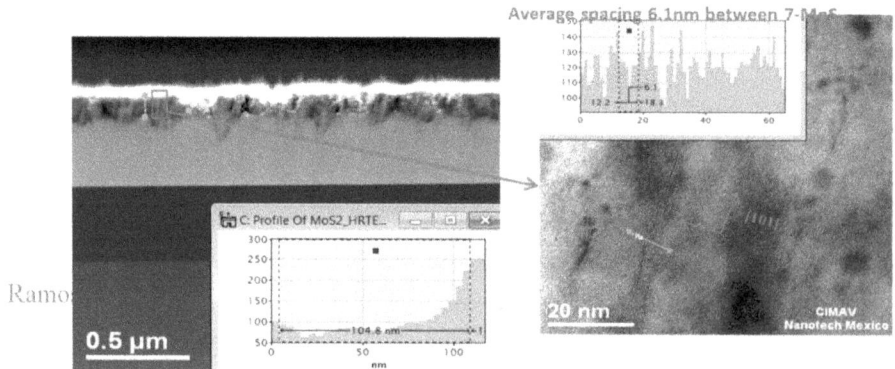

Figure 5.
High-resolution STEM image showing a vertical growth of MoS$_2$ crystallites as confirmed by 0.62 nm interlayer distance in (002) basal plane, in agreement. Image taken with rights and permissions from IOP-surf. Topogr.: Metrol. Prop.© Ramos et al. [22].

fused silica sample, under test parameters of C_0 = 24.06, C_1 = −184.31, C_2 = 6532.04, C_3 = −25482.45, and C_5 = 19015.30 as constant area of contact for continuous stiffness method (CSM) as described in detail by Li et al. [27]. All data was recorded by AFM Nano Vision© system attached to the nanoindenter system. The estimated values for hardness (H) and elastic modulus (E) were calculated using Eq. (2) to

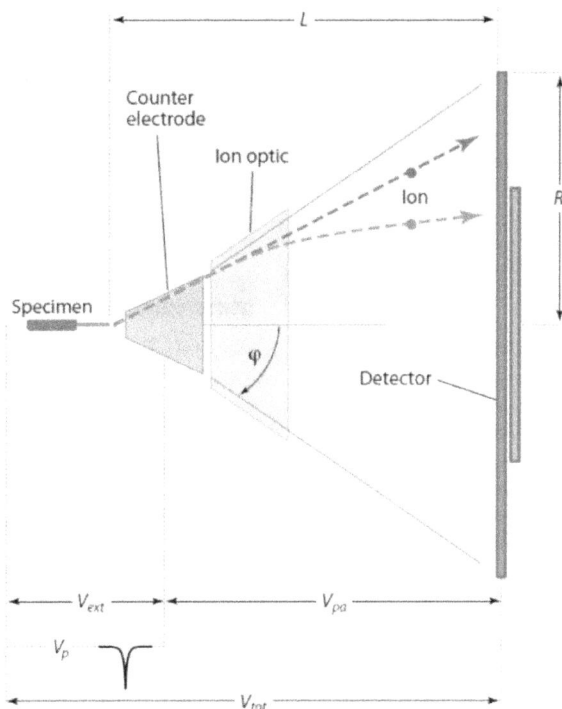

Figure 6.
Schematic drawing of APT (taken from [23]).

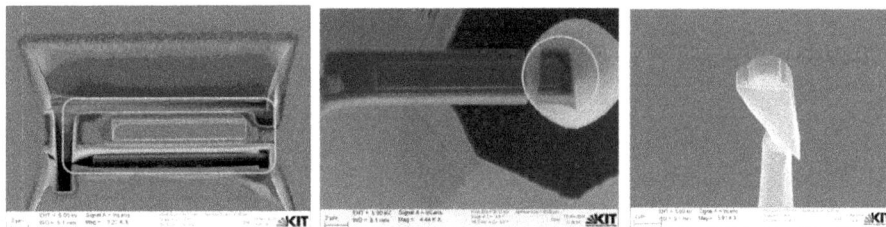

Figure 7.
Scanning electron images taken during lift-out procedure to prepare a needle for atom probe tomography in MoS₂ thin film (green square and circles are the areas of interest and cut using Ga ions and Omiprobe® micromanipulators), as discussed by Szász et al. [21].

determine stiffness S, when comparing to silicon substrates (001) surface termination and applying a continuous stiffness method as described extensively by Pharr et al. [28].

$$S = \left| \frac{1}{\frac{F_0}{Z_0}\cos\varphi - (K_s - m\,\omega^2)} - \frac{1}{K_f} \right|^{-1} \tag{2}$$

In Eq. (2), ω is the excitation frequency, (Z_0) displacement amplitude, (φ) phase angle, and (F_0) is the excitation amplitude, all those values can be obtained if the machine parameters load-frame stiffness K_f and stiffness of springs (K_s) as well the mass m are known input values during nanoindentation test. The coating hardness of film H_f can be estimated using a work indentation model described by

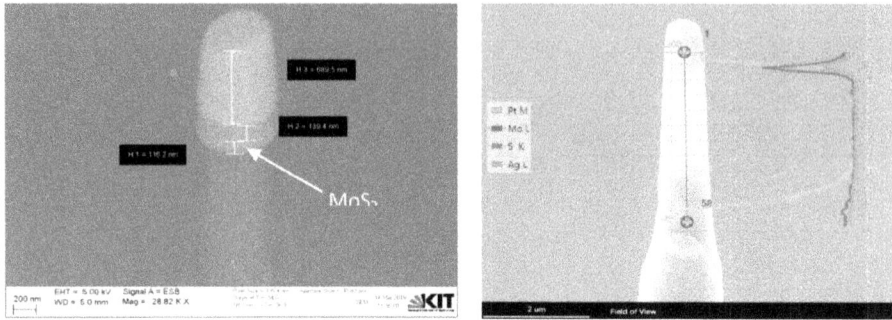

Figure 8.
Scanning electron images and line scan EDS to map chemical composition on the needle; molybdenum and sulfur atoms were detected over MoS₂ section (~ 110 nm).

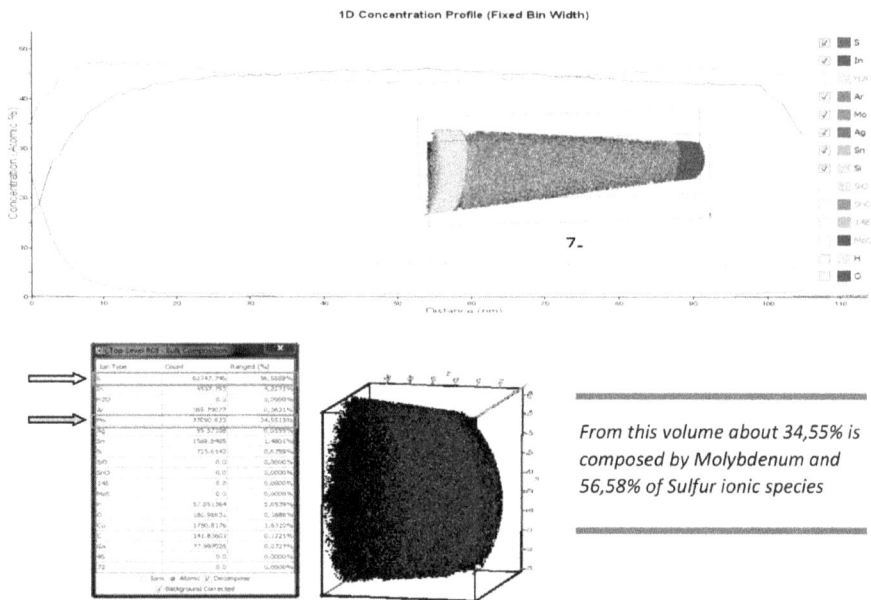

From this volume about 34,55% is composed by Molybdenum and 56,58% of Sulfur ionic species

Figure 9.
Top: *Atomic concentration profile from atom probe tomography for the MoS₂ thin film (~110 nm).* **Bottom:** *Portion of APT needle where corresponding to MoS₂ highest concentration as shown on ions table on left (yellow and blue arrows), in agreement with Singh et al. [18].*

Eq. (3), having **Hc** as composed film/substrate hardness; **Hs** and **Hf** as substrate and film hardness, the constant k represents a fitting parameter determined experimentally from the variation of **Hc** with relative indentation depth (β = **Hc/t**).

$$H_c = H_s + \frac{H_f - H_s}{1 + k\beta^2} \tag{3}$$

The elastic modulus E can be estimated using Eq. (4); having **Eeff** as effective reduced elastic modulus of the system in array film/substrate, contact area is determinate by **A** as function of the penetration depth, v is the Poisson ratio, t represents film thickness, and α is a parameter which depends on the material and the indenter geometry, in our case a pyramidal shape, as described by Domínguez-Rios et al. [29] and Hurtado-Macias et al. [30].

$$\frac{1}{E_{eff}} = \frac{(1 - v_f^2)}{E_f}(1 - e^{-at/\sqrt{A}}) + \frac{(1 - v_s^2)}{E_s}(e^{-at/\sqrt{A}}) + \frac{(1 - v_i^2)}{e_i} \qquad (4)$$

By using CSM method, it is was possible to estimate elastic modulus and hardness values as follows: Three regions of test are observed in the **Figures 12** and **13**, where region I is hardness values for MoS₂ crystallites with penetration depth of 0–90 nm, having no influence from silicon oxide substrate and a hardness value of $H = 6.0 \pm 0.1$ GPa and elastic modulus of $E = 136 \pm 2$ GPa. The region II, which has a penetration deep of 90–120 nm both values of elastic modulus and hardness are increased, meaning a clear influence by silicon oxide substrate, as confirmed by profilometry a thin film thickness of ~105 nm (both insets of **Figure 1**). The region III with penetration depth of 120–150 nm represents a hardness and elastic modulus of silicon oxide substrate, which are in partial agreement with Malzbender & With [31] to whom performed similar experiment on SiO₂ spin coated with methyltrimethoxysilane.

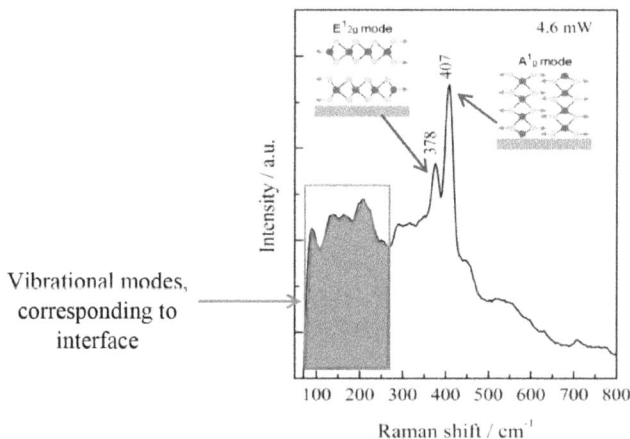

Figure 10.
The Raman spectra with two characteristic modes of vibrations at E$^1_{2g}$ at 378 cm^{-1} and A^1_g at 407 cm^{-1}, in agreement with Kong et al. [19].

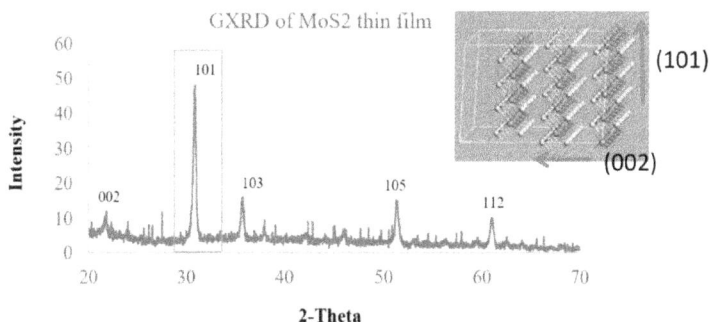

Figure 11.
Grazing incidence x-ray diffraction it was possible to observe a dominant (101) reflection at 2θ ~30°, in agreement with Liu et al. [26] for vertical aligned layers.

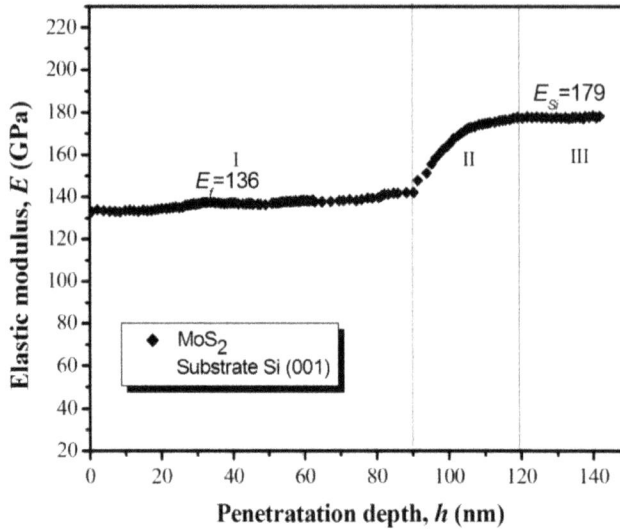

Figure 12.
Nanoindentation curves estimated experimentally using continuous stiffness method (CSM), the curve corresponds to regions I, II, III. In region I the estimated elastic modulus is **E** *= 136 ± 2 GPa corresponds to 0–90 nm of penetration depth, which is indicated to be only for MoS$_2$ film, in agreement with [28].*

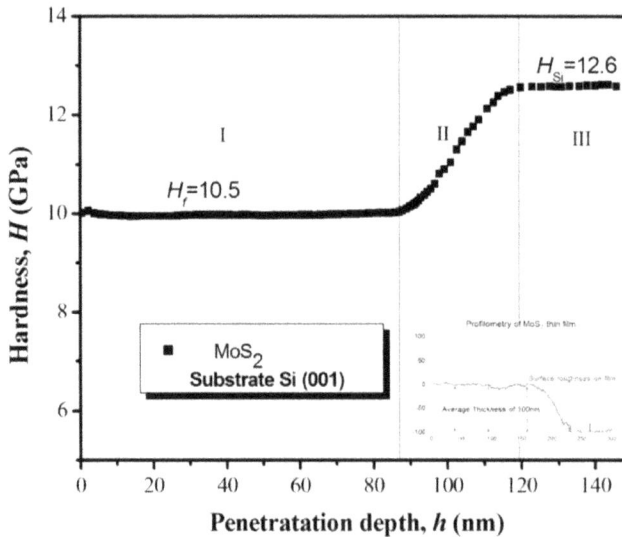

Figure 13.
The region I corresponds to hardness values of **H** *= 10.5 ± 0.1 GPa at 0–90 nm of penetration depth corresponding to MoS$_2$ layer. The regions II and III on both curves corresponds to mixed stage MoS$_2$/SiO$_2$ and SiO$_2$ substrate reason of an increase on both values are observed, in agreement with [28].*

The obtained values for hardness and elastic modulus are smaller estimations when comparing with results as presented by Bertolazzi et al. [12, 14] for single layers of MoS$_2$; we believe this occurs because of low dimension laminates can be stronger than stacking of MoS$_2$ crystallites. The applied force was done over (001)-basal plane as suspended on patterned silicon holes [12, 14], and in this case indenter tip can sweep MoS$_2$ crystallites over surface area. For that reason, our research team proceed to estimate film adherence by using AFM scratching technique in

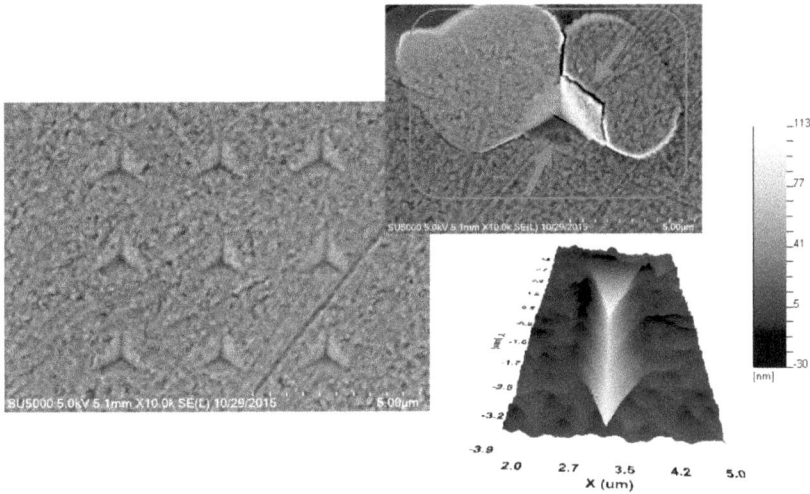

Figure 14.
Left: *Scanning electron micrograph indicating the nine zones of nanoindentation made with diamond indenter tip to estimate the elastic modulus and hardness values on MoS₂ film.* **Center:** *Scanning electron micrograph showing cracks over triangle shape indentation as indicated by red arrows.* **Right:** *Atomic force microscope starching zone to estimate a stiffness values of 4.27kN/m over the MoS₂ film.*

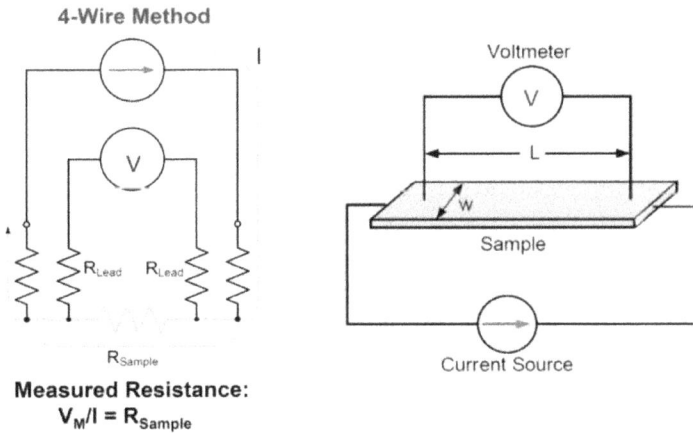

Figure 15.
Graphical description of the four-point method implemented for electrical transport in MoS₂ film.

encountering a deformation 0.85 μm² with a residual groove width 1 μm (a total groove height 125 nm and pile up height 40 nm), as presented in **Figure 14**, along with indentation sites completed to obtain elastic modulus and hardness values.

2.7 Electrical transport and resistivity

The electrical transport of the MoS₂ film matrix was investigated using four-point probe method as indicated in **Figure 15**, equipped with Keithley 4200-SCS in applied voltage range from −3.5 to 3.5 V. The transport measurements were done at room temperature and by direct contact to the MoS₂ film surface, no especial solder or metallic glue was used. Also, they were completed in the presence of light and dark conditions, the results indicate a linear Ohmic behavior, as presented in

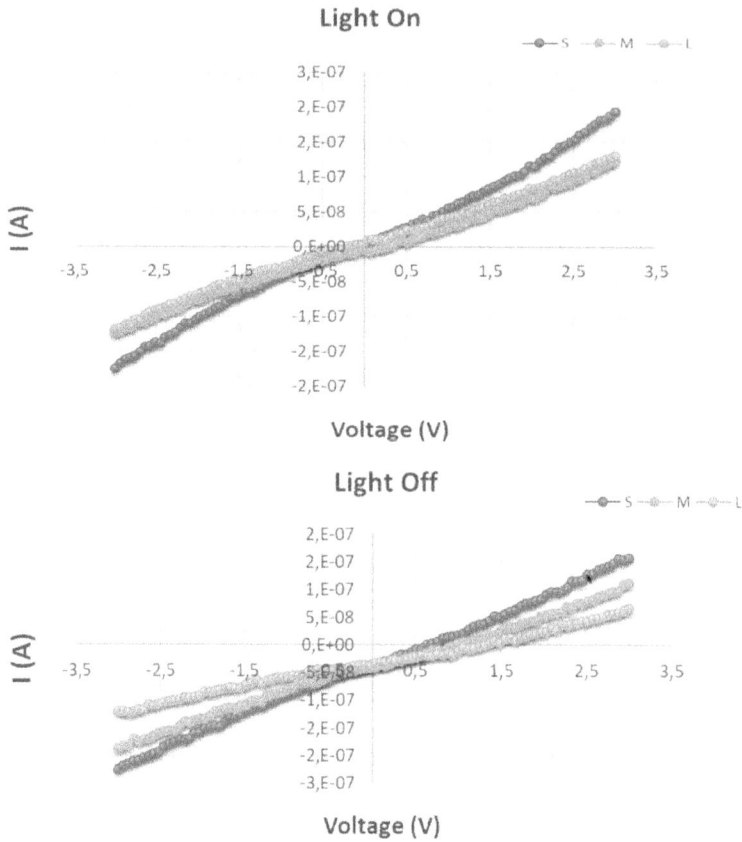

Light On

Light Off

Figure 16.
*Top: I-V curves measured in the presence of white light and using four-point method MoS₂ film. **Bottom:** I-V curves measured film in dark-room under otherwise conditions. When comparing both measurements it is possible to observe a change on slope, which is related to resistive values as presented in* **Figure 17**.

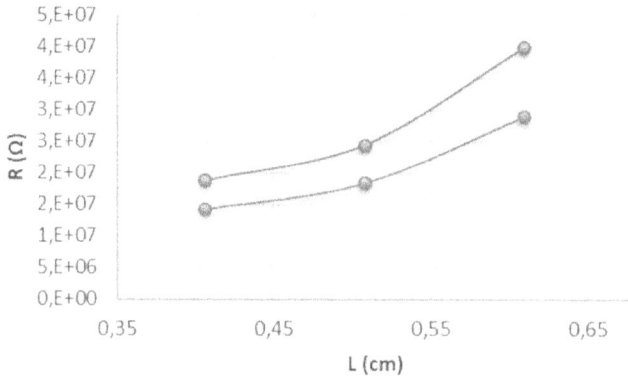

Figure 17.
Resistivity ρ values, calculated from I-V curves measured in the presence of white light and dark environment by using four-point method over MoS₂ film's surface. Red curve corresponds to presence of light shining on surface and blue curve to dark-room conditions.

Figure 16 and resistivity values in **Figure 17**, from **Tables 1** and **2** it was possible to determine values differences when white light is present, some authors refer this as photo-voltaic effect due to its intrinsic semiconductor nature of MoS_2 [2, 32, 33].

Size (m)	Impedance	Resistance	Power	Resistivity (Ω/m)
5×10^{-3}	0.4064	1.41×10^6	0.0381	26.507
10×10^{-3}	0.508	1.83×10^6	0.0381	27.473
15×10^{-3}	0.6096	2.89×10^6	0.0254	24.106

Table 1.
Values of impedance, resistance, power and resistivity measured without presence of white light.

Size (m)	Impedance	Resistance	Power	Resistivity (Ω/m)
5×10^{-3}	0.4064	1.86×10^6	0.0381	35.047
10×10^{-3}	0.508	2.46×10^6	0.0381	36.605
15×10^{-3}	0.6096	3.92×10^6	0.0254	33.256

Table 2.
Values of impedance, resistance, power and resistivity measured in presence of white light.

3. Discussion and conclusion

By using radio frequency sputtering techniques at high-vacuum it was possible to fabricate MoS_2 films with thickness of ~100 nm over pristine silicon oxide (SiO_2) wafers. The film surface analysis was carried out using electron microscopy and spectroscopy techniques and results indicate molybdenum di-sulfide had a vertical crystallite growth as shown in **Figures 2C** and **5**. Energy disperse confirms Sulfur-Kα (60%) and Molybdenum-Lα (40%) at 2.4 keV signal; and Raman spectroscopy modes of vibration at surface corresponding to E^1_{2g} = 378 cm^{-1} and A^1_g = 407 cm^{-1}. From high-resolution STEM it was possible to determine a degree of stacking between 7 layers along (002)-basal plane and to confirm vertical growth in agreement with Kong et al. [19], and APT preliminary measurements indicate a large quantity of sulfur and molybdenum with no grain boundaries or high impurities within film matrix for specific thin film growth using RF-sputtering conditions. From electrical transport measurements, it was possible to determine a linear Ohmic behavior and excitation when external visible light was on and off during four-point probe measurements as indicated by **Figure 16**, the resistivity values 26.5 Ω/m versus 35.0 Ω/m for off and on in external visible light, as possible caused by intrinsic semiconductor nature of MoS_2 in agreement with [2, 32, 33]. The mechanical properties were also investigated as previously reported by Ramos et al. [22] for indenter penetrating film surface 0-90 nm it was possible to estimate hardness of H = 10.5 ± 0.1 GPa and elastic modulus E = 136 ± 2 GPa by the continuous stiffness method [28]. It was concluded that MoS_2 films are a promising semiconducting material candidate for large scale photovoltaic applications, due to low cost and reliable and straight forward approach for homogenous low dimension films and its possible chemical combination with other group VI semiconducting materials as indicated by Najmaei et al. [34, 35] and Matis et al. [36] and Paranjape et al. [37–42].

Acknowledgements

The principal author thanks Universidad Autónoma de Ciudad Juárez for financial support by PIVA-2017 project titled: *"Fabricación de plantillas metálico-orgánicas: Una ruta hacia celdas solares más eficientes"* and PROFOCIE 2015. The

Laboratorio Nacional de Nanotecnología of Centro de Investigación en Materiales Avanzados (CIMAV-Chihuahua) and ICNAM program of Kleberg Advanced Laboratory Center of University of Texas at San Antonio both for the usage of electron microscopy and characterization equipment and facilities. Part of this work was performed at the Center for Integrated Nanotechnologies, an Office of Science User Facility operated for the U.S. Department of Energy (DOE) Office of Science. Sandia National Laboratories is a multi-program laboratory managed and operated by Sandia Corporation, a wholly owned subsidiary of Lockheed Martin Corporation, for the U.S. Department of Energy's National Nuclear Security Administration under contract DE-AC04-94AL85000. We are grateful to the Karlsruhe Micro and Nano Facility (KNMF) of Karlsruhe Institute of Technology for usage of Atom Probe Tomography and FIB instruments, under proposal 2018–020-022609. To facultad de ciencias de la electronica of Benemerita Universidad Autónoma de Puebla for usage of four-point probe station and laboratory facilities.

Conflict of interest

Authors declare no conflict of interest.

Author details

Manuel Ramos[1,2*], John Nogan[2], Manuela Ortíz-Díaz[1], José L Enriquez-Carrejo[1], Claudia A Rodriguez-González[1], José Mireles-Jr-Garcia[1], Roberto Carlos Ambrosio-Lazáro[4], Carlos Ornelas[3], Abel Hurtado-Macias[3], Torben Boll[5], Delphine Chassaing[5] and Martin Heilmaier[5]

1 Departamento de Física y Matemáticas, Instituto de Ingeniería y Tecnología, Universidad Autónoma de Cd. Juárez, Chihuahua, CP, Mexico

2 Center for Integrated Nanotechnologies, Albuquerque, NM, USA

3 Laboratorio Nacional de Nanotecnología, Centro de Investigación en Materiales Avanzados S.C., Chihuahua, CP, México

4 Facultad de Ciencias de la Electrónica, Benemérita Universidad Autónoma de Puebla, Ciudad de Puebla, Estado de Puebla, CP, México

5 Institut für Angewandte Materialien-Werkstoffkunde (IAM-WK), Karlsruher Institut für Technologie, Karlsruhe, Germany

*Address all correspondence to: manuel.ramos@uacj.mx

IntechOpen

References

[1] Chianelli RR, Berhault G, Torres B. Unsupported transition metal sulfide catalysts: 100 years of science and application. Catalysis Today. 2009;**147**:275-286

[2] Villareal A et al. Importance of the sulfidation step in the preparation of highly active NiMo/SiO$_2$/Al$_2$O$_3$ hydrodesulfurization catalysts. Catalysis Today. 2015;**250**:60-65

[3] Olivas A, Alonso G, Fuentes S. The catalytic activity of Ni/W bimetallic sulfide nanostructured catalysts in the hydrodesulfurization of dibenzothiophene. Topics in Catalysis. 2006;**39**:175-179

[4] Siadati MH, Alonso G, Torres B, Chianelli R. Open flow hot isostatic pressing assisted synthesis of unsupported MoS$_2$ catalysts. Applied Catalysis A. 2006;**305**:160-168

[5] Li W, Jun-fang C, HeQinyu WT. Electronic and elastic properties of MoS$_2$. Physica B. 2010;**405**:2498-2502

[6] Ramírez J, Macías G, Cedeño L, Gutiérrez-Alejandre A, Cuevas R, Castillo P. The role of titania in supported Mo, CoMo, NiMo, and NiW hydrodesulfurization catalysts: analysis of past and new evidences. Catalysis Today. 2004;**98**(1-2):19-30

[7] Stanislaus A, Marafi A, Rana MS. Recent advances in the science and technology of ultra low sulfur diesel (ULSD) production. Catalysis Today. 2010;**153**:1-68

[8] Morales-Ortuño JC, Ortega-Domínguez RA, Hernández-Hipólito O, Bokhimi X, Klimova TE. HDS performance of NiMo catalysts supported on nanostructured materials containing titania. Catalysis Today. 2016;**271**:127-139

[9] Gates BC et al. Catalysts tor emerging energy applications. MRS Bulletin. 2008;**33**:429-435

[10] Farragher AL, Cossee P. Proceedings of the 5th International Congress on Catalysis, North-Holland, Amsterdam; Vol. 1301. 1973

[11] Hagenbach G, Courty P, Delmon B. Physicochemical investigations and catalytic activity measurements on crystallized molydbenum sulfide-cobalt sulfide mixed catalysts. Journal of Catalysis. 1973;**31**:264-273

[12] Candia R et al. Proceedings of the 8th International Congress on Catalysis. Frankfurt-an-Main: Dechema; 1984;**2**:375

[13] Daage MM, Chianelli RR. Structure-function relations in molybdenum sulfide catalysts: The "Rim-Edge" model. Journal of Catalysis. 1994;**149**:414-427

[14] Perdew JP, Burke K, Ernzerhof M. Generalized Gradient Approximation Made Simple. Physical Review Letters. 1996;**77**:3865

[15] Hammer B, Hansen LB, Nørskov K. Improved adsorption energetics within density-functional theory using revised Perdew-Burke-Ernzerhof functionals. Physical Review B. 1999;**59**:7413

[16] Lauritsen JV et al. Location and coordination of promoter atoms in Co- and Ni-promoted MoS$_2$-based hydrotreating catalysts. Journal of Catalysis. 2007;**249**:220-233

[17] Remškar M, Viršek M, Mrzel A. The MoS$_2$ nanotube hybrids. Applied Physics Letters. 2009;**95**:133122

[18] Camacho-Bragado GA, Elechiguerra JL, Yacaman M. Characterization

of low dimensional molybdenum sulfide nanostructures. Materials Characterization. 2008;**59**:204-212

[19] Blanco E, Afanasiev P, Berhault G, Uzio D, Loridant S. Resonance Raman spectroscopy as a probe of the crystallite size of MoS_2 nanoparticles. Comptes Rendus Chimie. 2016;**19**:1310-1314

[20] Ramos MA et al. Spherical MoS_2 micro particles and their surface dispersion due to addition of cobalt promoters. Revista Mexicana de Física. 2011;**57**:220-223

[21] Ramos MA, Berhault G, Ferrer DA, Torres B, Chianelli RR. HRTEM and molecular modeling of the MoS_2–Co9S8 interface: understanding the promotion effect in bulk HDS catalysts. Catalysis Science & Technology. 2012;**2**:164-178

[22] Hansen LP, Johnson E, Brorson M, Helveg S. Growth mechanism for single- and multi-layer MoS_2 nanocrystals. Journal of Physical Chemistry C. 2014;**118**:22768-22773

[23] Casillas G et al. Elasticity of MoS_2 sheets by mechanical deformation observed by in situ electron microscopy. Journal of Physical Chemistry C. 2015;**119**:710-715

[24] Ramos M et al. In-situ HRTEM study of the reactive carbide phase of Co/MoS_2 catalyst. Ultramicroscopy. 2013;**127**:64-69

[25] Midgley PA, Dunin-Borkowski RE. Electron tomography and holography in materials science. Nature Materials. 2009;**8**:271-280

[26] Ziese U, de Jong KP, Koster AJ. Electron tomography: a tool for 3D structural probing of heterogeneous catalysts at the nanometer scale. Applied Catalysis A: General. 2004;**260**:71-74

[27] Arslan I, Marquis EA, Homer M, Hekmaty MA, Bartelt NC. Towards better 3-D reconstructions by combining electron tomography and atom-probe tomography. Ultramicroscopy. 2008;**108**:1579-1585

[28] Ma L, Chen W-X, Xu L-M, Zhou X-P, Jin B. One-pot hydrothermal synthesis of MoS_2 nanosheets/C hybrid microspheres. Ceramics International. 2012;**38**:229-234

[29] Sanders T, Prange M, Akatay C, Binev P. Physically motivated global alignment method for electron tomography. Advanced Structural and Chemical Imaging. 2015;**1**:11-11

[30] Sanders T. Discrete iterative partial segmentation technique (DIPS) for tomographic reconstruction. IEEE Transactions on Computational Imaging. 2016;**2**:71-82

[31] Coelho A. TOPAS-academic V4. Vol. 1. Australia: Coelho Software, Brisbane; 2007

[32] Magini M et al. Programme en FORTRAN IV pour l'analyse des données expérimentales relatives à la diffusion des rayons X par des substances liquides, amorphes et microcristallisées. Journal of Applied Crystallography. 1972;**5**:14

[33] Niemantsverdriet JW. et al. Catalysis—Biocatalytic Processes Encyclopedia of Life Support Systems (EOLSS). 2007

[34] Ertl G. Oscillatory catalytic reactions at single-crystal surfaces. Advances in Catalysis. 1990;**37**:213-277

[35] Imbihl R, Ertl G. Oscillatory Kinetics in Heterogeneous Catalysis. Chemical Reviews. 1995;**95**:697-733

[36] Wang S et al. A new molybdenum nitride catalyst with rhombohedral MoS_2 structure for hydrogenation applications. Journal of the American Chemical Society. 2015;**137**:4815-4822

[37] Galindo-Hernández F, Domínguez JM, Portales B. Structural and textural properties of Fe_2O_3/γ-Al_2O_3 catalysts and their importance in the catalytic reforming of CH_4 with H_2S for hydrogen production. Journal of Power Sources. 2015;**287**:13-24

[38] Samuel J, Ottolenghi M, Avnir D. Diffusion limited reactions at solid-liquid interfaces: Effects of surface geometry. Journal of Physical Chemistry. 1991;**95**:1890-1895

[39] Avnir D et al. Fractal analysis of size effects and surface morphology effects in catalysis and electrocatalysis. Chaos. 1991;**1**:397-410

[40] Magini M et al. Programme en FORTRAN IV pour l'analyse des données expérimentales relatives à la diffusion des rayons X par des substances liquides, amorphes et microcristallisées. Journal of Applied Crystallography. 1972;**5**:14

[41] Seri-Levy A, Avnir D. Effects of heterogeneous surface geometry on adsorption. Langmuir. 1993;**9**:3067-3076

[42] Neimark AV. Calculating Surface Fractal Dimensions of Adsorbents. Adsorption Science Technology. 1991;7:210-219

www.ingramcontent.com/pod-product-compliance
Lightning Source LLC
Chambersburg PA
CBHW070156240326
41458CB00127B/5824